An Invitation to the
ROGERS-RAMANUJAN
IDENTITIES

An Invitation to the
ROGERS-RAMANUJAN
IDENTITIES

Andrew V. Sills

CRC Press
Taylor & Francis Group
Boca Raton London New York

CRC Press is an imprint of the
Taylor & Francis Group, an **informa** business
A CHAPMAN & HALL BOOK

The only known photograph of L. J. Rogers appears on the cover and on page xv. This photo appeared, undated and uncredited, in Rogers' Royal Society obituary in December 1934.

The image of S. Ramanujan appearing on the cover and on page xvi is from one of four extant photographs of Ramanujan according to Bruce Berndt, "The Four Photographs of Ramanujan," *Ramanujan: Essays and Surveys*, ed. B. C. Berndt and R. A. Rankin, in: *History of Mathematics*, vol. 22, American Mathematical Society and London Mathematical Society, 2001. The photographer is unknown, but according to Berndt, the photograph was likely taken during the summer of 1916.

The publisher believes both images to be in the public domain.

CRC Press
Taylor & Francis Group
6000 Broken Sound Parkway NW, Suite 300
Boca Raton, FL 33487-2742

First issued in paperback 2020

© 2018 by Taylor & Francis Group, LLC
CRC Press is an imprint of Taylor & Francis Group, an Informa business

No claim to original U.S. Government works

ISBN-13: 978-1-4987-4525-3 (hbk)
ISBN-13: 978-0-367-65761-1 (pbk)

**Visit the Taylor & Francis Web site at
http://www.taylorandfrancis.com**

**and the CRC Press Web site at
http://www.crcpress.com**

To my family

Contents

Contents

Foreword

I first met the Rogers–Ramanujan identities in the fall semester of 1961 at the University of Pennsylvania. The course was taught by Hans Rademacher, one of the great number theorists of the twentieth century. He clearly was in agreement with Hardy's comment that "[i]t would be difficult to find more beautiful formulæ than the 'Rogers–Ramanujan' identities" [Ram27, p. xxxiv]. Rather poignantly, he believed that D. H. Lehmer had proved that these two identities were an isolated phenomenon. While this belief later turned out to be false, it is clear that Rademacher found these results to be so beautiful that it was important to reveal such esthetic excellence to beginning graduate students.

Rademacher's pessimistic view of the Rogers–Ramanujan identities as beautiful but singular results turned out to be far from the truth.

The recognition of the Rogers–Ramanujan identities as the tip of a research cornucopia is surely attributable to Basil Gordon's generalization in 1961 [Gor61]. Rademacher was completely unaware of Gordon's paper, and, consequently, so were all of his students (including me). I would note that in "Some Debts I Owe" [And01], I mention how Rademacher's proof of the Rogers–Ramanujan identities led me to an independent discovery of Gordon's theorem. I was truly deflated when I learned of Gordon's work months after I had submitted this grand generalization for publication.

To make a long story short, I was completely captivated by the Rogers–Ramanujun identities after Rademacher's enticing reënactment of Schur's proof. Since then they have been a constant theme in much of my research.

Now Andrew (Drew) Sills has put together this great introduction to these fascinating results. Drew was my student and wrote his PhD thesis under my direction, elucidating and extending the work of Lucy Slater on a variety of q-series identities of Rogers–Ramanujan type.

This is a marvelous book. Drew has drawn on his encyclopedic knowledge of the literature to prepare this coherent and exciting account of the Rogers–Ramanujan identities and the aftermath. The book has numerous exercises and so could well be used as a text in a graduate course. Drew has clearly in mind that this book is an introduction. So, more advanced topics are saved to the final "But wait... there's more" chapter where pointers are given to sources in the literature.

But wait... there's even more! Appendix A provides an extended version of Lucy Slater's list, plus an extended list of false theta function identities, and Appendix B contains some of the wonderful letters written by the early pioneers of the subject.

This is wonderful. Thank you, Drew, for this excellent book!

George E. Andrews
Evan Pugh University Professor in Mathematics
Pennsylvania State University
University Park, PA 16802

Preface

Leonard James Rogers (1862–1933)

In 1894, a relatively unknown English mathematician at Yorkshire College named Leonard James Rogers published a paper entitled "Second memoir on the expansion of certain infinite products," in the *Proceedings of the London Mathematical Society* [Rog94]. This second installment of a three-part series on the expansion of infinite products was 26 pages long. Buried on the tenth page of this paper, and written in obscure notation, is an identity of analytic functions, valid when $|q| < 1$, that later came to be known as the first Rogers–Ramanujan identity [Rog94, p. 328]:

$$\sum_{n=0}^{\infty} \frac{q^{n^2}}{(1-q)(1-q^2)\cdots(1-q^n)} = \prod_{k=0}^{\infty} \frac{1}{(1-q^{5k+1})(1-q^{5k+4})}. \qquad (0.1)$$

What we now know as the second Rogers–Ramanujan identity occurs two pages later [Rog94, p. 330]:

$$\sum_{n=0}^{\infty} \frac{q^{n^2+n}}{(1-q)(1-q^2)\cdots(1-q^n)} = \prod_{k=0}^{\infty} \frac{1}{(1-q^{5k+2})(1-q^{5k+3})}. \qquad (0.2)$$

G. H. Hardy's account of how (0.1) and (0.2) were discovered by Rogers, ignored by the mathematical community, only to be rediscovered (conjecturally) many years later by Srinivasa Ramanujan is quoted in Chapter 2.

Srinivasa Ramanujan (1887–1920)

The uninitiated will naturally wonder how any author could purport to write an entire book about nothing more than a pair of identities of analytic functions. In contrast, experts in the field will realize immediately upon picking up this volume that it cannot possibly be long enough to adequately discuss the multitude of subjects arising from (0.1) and (0.2). In the interest of full disclosure, yes, this volume is but an *introduction* to *certain aspects* of the Rogers–Ramanujan identities, and an invitation to study further.

The Rogers–Ramanujan identities, and identities of similar type, arise in number theory, analysis, combinatorics, the theory of integer partitions, vertex operator algebras, representation theory of Lie algebras, statistical mechanics, and knot theory. A goal of this book is to present and develop the Rogers–Ramanujan identities from the perspective of the theory of basic hypergeometric series and the theory of integer partitions. Historical material, in the form of remarks, and in some cases longer quotes and explanations, is woven into the narrative. While I have in some sense attempted, in broad terms, to present the material in a kind of "historical arc," I have, for mathematical efficiency and expediency, introduced modern techniques as required. For example, I begin with a "prehistory" chapter where integer partitions and hypergeometric series are introduced in order to set the stage for the unveiling of the Rogers–Ramanujan identities in the next chapter. However, in order to deal with hypergeometric series and their generalizations efficiently, I introduce the techniques of Wilf and Zeilberger, which of course made their debut nearly a hundred years *after* the discovery of the Rogers–Ramanujan identities. This is done so for two reasons: to convey how modern practitioners approach (*q*-)hypergeometric series, and so that a plethora of useful classical results can be dispensed with via one-line proofs.

In the course of this work, I take the liberty of not attempting to make the narrative entirely "self-contained" in the sense that I mention related topics and results of interest, even if I do not have the space to fully develop the topic and give proofs of those results. To this end, one of the purposes of this book is to alert interested readers to what related material is in the literature and guide them to that literature. A case in point is the Rademacher convergent series for the unrestricted partition function, and related results for various restricted partition functions. A number of my students over the years were instantly enthralled with these theorems and were led to study the circle method as a result. Accordingly, I mention them in this book, but as excellent expositions of the circle method are readily available elsewhere in the literature, I direct the readers to those sources rather than providing a similar development here.

In another attempt to engage readers who may be interested in the lives of some of the people who are responsible for the mathematics discussed in this book, I have included in an appendix transcriptions of letters written by W. N. Bailey to Freeman Dyson and to Lucy Slater. Readers will gain some "inside" information on how the mathematics is actually *done*, before it was ready to be written up as the polished final product that we read in the journals.

The mathematical prerequisites assumed are fairly minor. In some sense, not much beyond an elementary calculus course and an introductory number theory course is required to follow the statements and proofs of the theorems included. On the other hand, a certain level of mathematical experience and sophistication is required to fully appreciate the material presented. For instance, I have found that in most cases, students who have recently completed a year of calculus, and thus have some experience with power series, are nonetheless not quite ready to work with generating functions. The first two chapters are long and contain many exercises. These two chapters alone could easily form the material of a special topics graduate course. Chapters 3–5 contain more advanced material, and fewer explicitly stated exercises, as graduate students and practitioners at this level will have the sophistication to ask themselves relevant questions, and practice as needed to master the material to their own satisfaction. Chapter 6 contains eight sections, each of which, if more fully developed, could have been a chapter in its own right. The interested reader is pointed to the literature for more information.

All that having been stated, I hope that there is plenty contained in this volume to engage both graduate students and strong undergraduates, as well as material that professional mathematicians and scientists will find both useful and delightful.

Andrew Sills
Savannah, Georgia
June 2017

Acknowledgments

I have many people to thank for helping this book come into being.

In particular, this project began as a result of Sarfraz Kahn of CRC Press asking to meet with me at the Joint Meetings of the American Mathematical Society and the Mathematical Association of America held in San Antonio, Texas, in January, 2015. He has been supportive and encouraging throughout the entire process. I also wish to express my gratitude to Project Editor Robin Lloyd-Starkes and her team for all their help, and to "TEX-pert" Shashi Kumar for assistance with typesetting issues.

I thank George Andrews for being my teacher, thesis advisor, mentor, and friend over many years. George first introduced me to the Rogers–Ramanujan identities, and clearly this book would have never come into being without his guidance and inspiring influence. I also thank George for granting me access to his files, including the letters W. N. Bailey wrote to Lucy Slater. I am also grateful for his writing the *Foreword* to this book.

I am grateful to Freeman Dyson for preserving the letters he received from W. N. Bailey in the 1940s and for allowing me to transcribe and present them in this book.

I thank Mike Hirschhorn for much stimulating conversation and correspondence over the years, and in particular for pointing out that Marshall Hall's notes on Hardy's 1936 IAS lectures [Har37b] contain a statement that Ramanujan had in fact found a proof of the Rogers–Ramanujan identities in 1917 shortly before Ramanujan stumbled upon Rogers' published proof in the *Proceedings of the London Mathematical Society*.

Thanks are due to Jim Lepowsky, who read an early version of the manuscript and patiently shared his mathematical insights and helped with the wording of a number of sections, especially those which dealt with the intersection of Rogers–Ramanujan type identities with vertex operator algebra theory. Any misstatements or questionable wording that remain are entirely the fault of the author, and remain despite Jim's painstaking efforts.

I thank Ken Ono for sharing thoughts on how best to present the material in §6.5.

Many thanks to Jimmy McLaughlin and Peter Zimmer for collaborating with me on a number of research projects over the years, some of which are mentioned in this book.

I extend my sincere gratitude to Robert Schneider, who graciously volunteered to proofread the entire book before it went to press, and made numerous helpful suggestions.

I appreciate Doron Zeilberger for mentoring me during my postdoctoral years, and for continuing to encourage and collaborate with me since that time. Doron's influence will be felt throughout the book. In addition to the obvious use of his algorithms for efficiently proving hypergeometric and basic hypergeometric identities, his preference that the omnipresent variable q be treated as a formal variable in every instance when it is possible to do so, has been honored.

Others that certainly need to be acknowledged include Krishnaswami Alladi, Richard Askey, Alex Berkovich, Bruce Berndt, Doug Bowman, David Bressoud, Stefano Capparelli, Yuriy Choliy, Sylvie Corteel, Dennis Eichhorn, Frank Garvan, Ira Gessel, Mourad Ismail, Shashank Kanade, Karl Mahlburg, Steve Milne, Debajyoti Nandi, Peter Paule, Helmut Prodinger, Matthew Russell, José Plíno de Oliveira Santos, Carla Savage, James Sellers, Nic Smoot, Michael Somos, Dennis Stanton, Ole Warnaar, the late Herbert S. Wilf, Robert Wilson, and Ae Ja Yee.

And, of course, I thank my wonderful family for being my support network and my *raison d'être*.

Symbol Description

$\#S$	cardinality of the set S	$\|\lambda\|$	weight of (sum of parts in) the partition λ
$(a)_n$	rising factorial, Pochhammer symbol	$p(n)$	number of partitions of the integer n
$(a;q)_n$	rising q-factorial, q-Pochhammer symbol, q-shifted factorial	$p_l(n)$	number of partitions of n of length at most l
$[\alpha]_q$	$\frac{1-q^\alpha}{1-q}$; the basic analog of the number α	$p(l,n)$	number of partitions of n of length exactly l
crm_n	circulator (discrete periodic) function modulo m	$pd(n)$	number of partitions of n into distinct parts
D	ordinary differentiation operator	$po(n)$	number of partitions of n into odd parts
D_q	q-derivative operator	$\Re$	real part of a complex number
$E_q(z)$	a q-analog of e^z	$\binom{n}{k}$	binomial coëfficient
$e_q(z)$	another q-analog of e^z	$\binom{n}{k}_q$	q-binomial coëfficient, Gaussian polynomial
$_rF_s$	hypergeometric series	$\left(\!\binom{n}{k}\!\right)$	trinomial coëfficient
$_r\phi_s$	basic hypergeometric series, q-hypergeometric series	$\left(\!\binom{n}{k}\!\right)_q$	q-trinomial coëfficient of the first type
$f(a,b)$	Ramanujan's theta function	T_0	another q-trinomial coëfficient
$\vartheta(z,w)$	Jacobi's theta function	T_1	yet another q-trinomial coëfficient
$\varphi(q)$	$f(q,q)$	$U(m,A;q)$	$T_0(m,A;q) + T_0(m + 1, A; q)$
$\psi(q)$	$f(q,q^3)$		
$\Psi(a,b)$	false theta function	$[\![P]\!]$	Iverson bracket: 1 of P is true and 0 if P is false
λ_i	ith largest part of the partition λ		
λ'	conjugate of the partition λ	$\mathbb{Z}_k^\times$	The multiplicative group of units in $\mathbb{Z}_k$, i.e. $\{h : 0 \leq h < k, \gcd(h,k) = 1\}$
$\ell(\lambda)$	length of (number of parts in) the partition λ		

Chapter 1

Background and the Pre-History

"Read Euler, read Euler, he's the master of us all."—Pièrre-Simon
Laplace.

1.1 Elementary Theory of Partitions

Euler received a letter from Philip Naudé in September 1740 in which he
was asked how many ways one could write 50 as a sum of different positive
integers. This stimulus was sufficient to move Euler to inaugurate the theory
of partitions.

1.1.1 Partitions and Generating Functions

Definition 1.1. A *partition* λ of an integer n is a finite sequence
$(\lambda_1, \lambda_2, \ldots, \lambda_l)$ of positive integers (called the *parts* of λ) where

$$\lambda_1 \geq \lambda_2 \geq \cdots \geq \lambda_l$$

and

$$\lambda_1 + \lambda_2 + \cdots + \lambda_l = n.$$

The number of parts $l = \ell(\lambda)$ of λ is the *length* of λ.

We shall use the convention that λ_i is the ith part of the partition λ. It
will be convenient to define $\lambda_i := 0$ whenever $i > \ell(\lambda)$.

Example 1.2. The partitions of 5 are

$$(5) \quad (4,1) \quad (3,2) \quad (3,1,1) \quad (2,2,1) \quad (2,1,1,1) \quad (1,1,1,1,1).$$

If no part in a partition exceeds 9, it may be convenient to omit the commas
between parts, so that the seven partitions of 5 may be written

$$(5) \quad (41) \quad (32) \quad (311) \quad (211) \quad (2111) \quad (11111).$$

Some further definitions connected with partitions follow.

1

Definition 1.3. If λ is a partition of n, we also call n the *weight* of λ and may denote the weight of λ by $|\lambda|$. The number of times the integer j appears as a part in the partition λ is called the *multiplicity of j in λ* and is denoted $m_j = m_j(\lambda)$.

Another way to represent a partition is via the "superscript–multiplicity notation":

$$\langle 1^{m_1} 2^{m_2} 3^{m_3} \cdots \rangle,$$

where we may omit i^{m_i} if $m_i = 0$ and we may omit the superscripts m_i that equal 1. Thus the seven partitions of 5 may be written

$$\langle 5 \rangle, \quad \langle 1\ 4 \rangle, \quad \langle 2\ 3 \rangle, \quad \langle 1^2 3 \rangle, \quad \langle 1\ 2^2 \rangle, \quad \langle 1^3 2 \rangle, \quad \langle 1^5 \rangle.$$

Definition 1.4. The *sum* $\lambda + \mu$ of two partitions λ and μ is given by

$$\lambda + \mu = (\lambda_1 + \mu_1, \lambda_2 + \mu_2, \ldots, \lambda_M + \mu_M),$$

where $M = \max(\ell(\lambda), \ell(\mu))$.

Euler's theory of partitions begins with the following sequence of observations.

Consider the product

$$\frac{1}{1-q} \times \frac{1}{1-q^2}. \tag{1.1}$$

If we expand each factor as a geometric series

$$(1 + q^1 + q^{1+1} + q^{1+1+1} + \ldots) \times (1 + q^2 + q^{2+2} + q^{2+2+2} + \ldots) \tag{1.2}$$

and take care not to "simplify" the exponents, we obtain

$$1 + q^1 + (q^2 + q^{1+1}) + (q^{2+1} + q^{1+1+1}) + (q^{2+2} + q^{2+1+1} + q^{1+1+1+1}) + \cdots,$$

clearly demonstrating that all possible partitions employing only 1s and 2s as parts are generated in the exponents of the indeterminate q.

Next, notice that the coëfficient of q^n in the expansion of (1.1) is the number of partitions of n where 1s and 2s are the only allowable parts.

By identical reasoning,

$$\sum_{n=0}^{\infty} p_m(n)q^n = \prod_{k=1}^{m} \frac{1}{1-q^k},$$

where $p_m(n)$ denotes the number of partitions of n into parts no greater than m. Letting $m \to \infty$, we obtain Euler's generating function for the partition counting function:

Theorem 1.5 (Euler's generating function for $p(n)$).

$$\sum_{n=0}^{\infty} p(n)q^n = \prod_{k=1}^{\infty} \frac{1}{1-q^k}, \tag{1.3}$$

where $p(n)$ denotes the number of partitions of n.

Note that the constant term in the expansion of (1.3) is 1, and thus we define $p(0) := 1$. This definition may be further justified by considering the empty partition () to be the unique partition of 0, and the only partition of length zero. Further, we note that $p(n) = 0$ for all $n < 0$.

Euler's generating function for $p(n)$ is just the first of many generating functions to be considered, so let us make a formal definition.

Definition 1.6. A function $F(q)$ is called the *generating function* of the arithmetic function $f(n)$ if

$$F(q) = \sum_{n=0}^{\infty} f(n)q^n.$$

A classical method for proving identities between various classes of restricted partitions is via algebraic manipulation of formal infinite products.

A partition λ into distinct parts is one in which $m_j(\lambda) \leqq 1$ for all j. For example, the partitions of 9 into distinct parts are

$$(9) \quad (81) \quad (72) \quad (63) \quad (621) \quad (54) \quad (531) \quad (432);$$

eight in all. Notice that the partitions of 9 into odd parts are

$$(9) \quad (711) \quad (531) \quad (51111) \quad (333) \quad (33111) \quad (3111111) \quad (111111111);$$

again eight in all. It is not a coincidence that there are the same number of partitions of 9 into distinct parts as there are partitions of 9 into odd parts. Euler showed that this is true for all n, as proved in the next theorem.

Remark 1.7. Throughout this work, the symbol q (and any auxiliary variables that are needed) shall be viewed as formal variables, and the series and infinite products in which these variables appear shall be formal power series and formal infinite products respectively. In the algebraic manipulations performed (such as in the proof of Theorem 1.8 below), the notion of analytic convergence is unnecessary and would actually be a distraction from the goals to be achieved. However, it is the case that all of the q-series identities presented herein are also analytic identities[1], convergent when $|q| < 1$, but this will not be the viewpoint taken here. On those occasions when we temporarily go wading in analytic waters, and q needs to be regarded as a complex variable, this will be explicitly noted in the text.

Theorem 1.8 (Euler's partition theorem). *For all integers n, the number of partitions of n into odd parts equals the number of partitions of n into distinct parts.*

---

[1]See Historical Note 2.35!

Proof. Let $pd(n)$ denote the number of partitions of n into distinct parts. Let $po(n)$ denote the number of partitions of n into odd parts.

$$\sum_{n \geq 0} pd(n)q^n = \prod_{k \geq 1}(1 + q^k)$$

$$= \prod_{k \geq 1} \frac{1 - q^{2k}}{1 - q^k} \quad \left(\text{since } (1 - x)(1 + x) = (1 - x^2)\right)$$

$$= \prod_{k \geq 1} \frac{1 - q^{2k}}{(1 - q^{2k-1})(1 - q^{2k})}$$

$$= \prod_{k \geq 1} \frac{1}{1 - q^{2k-1}}$$

$$= \sum_{n \geq 0} po(n)q^n.$$

The result follows from comparing the coëfficient of q^n in the extremes. $\square$

Euler's partition theorem is interesting in and of itself, but perhaps more importantly, it demonstrates that any time we can find an equality of infinite products and interpret each side as a generating function for a certain restricted class of partitions, we have a partition identity.

For example, observe that

$$\prod_{j=0}^{\infty}(1 + q^{3j+1})(1 + q^{3j+2}) = \prod_{j=0}^{\infty} \frac{(1 - q^{6j+2})(1 - q^{6j+4})}{(1 - q^{3j+1})(1 - q^{3j+2})}$$

$$= \prod_{j=0}^{\infty} \frac{(1 - q^{6j+2})(1 - q^{6j+4})}{(1 - q^{6j+1})(1 - q^{6j+4})(1 - q^{6j+2})(1 - q^{6j+5})}$$

$$= \prod_{j=0}^{\infty} \frac{1}{(1 - q^{6j+1})(1 - q^{6j+5})}.$$

Next, observe that

$$\prod_{j=0}^{\infty}(1 + q^{2j+1} + q^{(2j+1)+(2j+1)}) = \prod_{j=0}^{\infty}(1 + q^{2j+1} + q^{4j+2})$$

$$= \prod_{j=0}^{\infty} \frac{1 - q^{6j+3}}{1 - q^{2j+1}}$$

$$= \prod_{j=0}^{\infty} \frac{1 - q^{6j+3}}{(1 - q^{6j+1})(1 - q^{6j+3})(1 - q^{6j+5})}$$

$$= \prod_{j=0}^{\infty} \frac{1}{(1 - q^{6j+1})(1 - q^{6j+5})}.$$

Thus,

$$\prod_{j=0}^{\infty}(1+q^{3j+1})(1+q^{3j+2}) = \prod_{j=0}^{\infty}(1+q^{2j+1}+q^{(2j+1)+(2j+1)})$$

$$= \prod_{j=0}^{\infty} \frac{1}{(1-q^{6j+1})(1-q^{6j+5})},$$

and we may conclude the truth of the following theorem.

Theorem 1.9. *Let $A(n)$ denote number of partitions of n into distinct non-multiples of 3. Let $B(n)$ denote the number of partitions of n wherein all parts are odd and no part may appear more than twice. Let $C(n)$ denote the number of partitions of n into parts congruent to ± 1 (mod 6). Then $A(n) = B(n) = C(n)$ for all n.*

Remark 1.10. In Theorem 1.9, the equality $A(n) = C(n)$ is the easy part of a theorem of Issai Schur from 1926, that we shall encounter in the next chapter. The fact that $B(n)$ also equals $C(n)$ and $A(n)$, although elementary, appears to have gone unnoticed until the late 1990s when it was observed by Krishnaswami Alladi and communicated privately to George Andrews.

1.1.2 Graphical and Combinatorial Methods

After Euler, the theory of partitions did not see many significant advances until James Joseph Sylvester became interested in the subject in the 1880s. Sylvester, together with some of his students and colleagues made many fundamental contributions to the theory of partitions. The simple, yet surprisingly effective device of representing a partition graphically as a so-called "Ferrers diagram" was named after Norman Macleod Ferrers, a British mathematician who edited the *Quarterly Journal of Pure and Applied Mathematics* with Sylvester and others during the second half of the nineteenth century.

Let π denote a partition of weight n and length k. The Ferrers diagram of π consists of a total of n dots in k left-justified rows, where the number of dots in row i corresponds to the i-th largest part of π.

For example the partition (5542111) has the following Ferrers diagram:

Definition 1.11. The *conjugate* of a partition λ, denoted λ', is the partition obtained from λ by reflecting the Ferrers diagram of λ along its main diagonal, i.e., by interchanging the rows and columns of the Ferrers diagram of λ.

The conjugate of the partition (5542111) is therefore (74332).

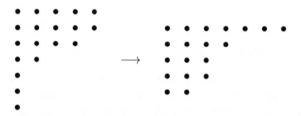

Notice that the Ferrers diagram of a partition of n with k parts, and largest part l, may be visualized as a $k \times l$ matrix, in which n of the entries equal 1 (with 1s corresponding to dots) and the rest of the entries equal 00 (with 0s corresponding to blanks in the Ferrers diagram). Then the Ferrers diagram of the conjugate λ' of a partition λ is simply the matrix transpose of the Ferrers diagram of λ.

$$
\begin{bmatrix}
1 & 1 & 1 & 1 & 1 \\
1 & 1 & 1 & 1 & 1 \\
1 & 1 & 1 & 1 & 0 \\
1 & 1 & 0 & 0 & 0 \\
1 & 0 & 0 & 0 & 0 \\
1 & 0 & 0 & 0 & 0 \\
1 & 0 & 0 & 0 & 0
\end{bmatrix}
\longrightarrow
\begin{bmatrix}
1 & 1 & 1 & 1 & 1 & 1 & 1 \\
1 & 1 & 1 & 1 & 0 & 0 & 0 \\
1 & 1 & 1 & 0 & 0 & 0 & 0 \\
1 & 1 & 1 & 0 & 0 & 0 & 0 \\
1 & 1 & 0 & 0 & 0 & 0 & 0
\end{bmatrix}
$$

The number of rows in the Ferrers diagram of a partition λ is the number of parts in λ, while the number of columns is the largest part of λ. Thus the following result is immediate:

Theorem 1.12. *The number of partitions of n with m parts equals the number of partitions of n with largest part m.*

Proof. Conjugation supplies a bijection between the two classes of partitions.
□

Definition 1.13. A partition λ is called *self-conjugate* if $\lambda' = \lambda$.

For example, $(55222)' = (55222)$

Remark 1.14. We could equivalently define the conjugate $\lambda' = (\lambda'_1, \lambda'_2, \ldots, \lambda'_{\lambda_1})$ of $\lambda = (\lambda_1, \lambda_2, \ldots, \lambda_{\ell(\lambda)})$ as that partition in which the ith part is

$$
\lambda'_i = \sum_{k=i}^{\lambda_1} m_k(\lambda)
$$

for $i = 1, 2, \ldots, \lambda_1$, or equivalently,

$$\lambda' = \langle 1^{\lambda_1 - \lambda_2} 2^{\lambda_2 - \lambda_3} 3^{\lambda_3 - \lambda_4} 4^{\lambda_4 - \lambda_5} \cdots \rangle. \tag{1.4}$$

For our next result, it will be convenient to have a special notation for partitions in which it is immediately obvious when two partitions are conjugate.

Definition 1.15. The *Durfee square* of a partition λ is the largest square of dots contained in the upper left corner of the Ferrers diagram of λ. The side length (or equivalently diagonal length) of a Durfee square is called its *order*.

For example, the partition (5542111) has a Durfee square of order 3.

Historical Note 1.16. William P. Durfee (1855–1941) was an American mathematician who earned his PhD in 1883 under the supervision of Sylvester at Johns Hopkins University. He joined the faculty of Hobart College in Geneva, New York, in 1884, where he was appointed dean in 1888.

Definition 1.17. Suppose λ is a partition of n, with a Durfee square of order r. Then the *Frobenius symbol* of λ is

$$\begin{pmatrix} a_1 & a_2 & \cdots & a_r \\ b_1 & b_2 & \cdots & b_r \end{pmatrix},$$

where $a_i = \lambda_i - i$ and $b_i = \lambda_i' - i$.

Notice that $a_1 > a_2 > \cdots > a_r \geq 0$ and $b_1 > b_2 > \cdots > b_r \geq 0$, and $n = r + \sum_{i=1}^{r}(a_i + b_i)$, and that the conjugate is obtained by simply swapping the two rows.

For example, the Frobenius symbol of (5542111) is

$$\begin{pmatrix} 4 & 3 & 1 \\ 6 & 2 & 0 \end{pmatrix},$$

while its conjugate partition, (74332), has Frobenius symbol

$$\begin{pmatrix} 6 & 2 & 0 \\ 4 & 3 & 1 \end{pmatrix}.$$

Theorem 1.18 (Sylvester). *The number of self-conjugate partitions of n equals the number of partitions of n into distinct odd parts.*

Proof. Let π be a self-conjugate partition of n with Durfee square of order r. Then π has Frobenius symbol

$$\begin{pmatrix} a_1 & a_2 & \cdots & a_r \\ a_1 & a_2 & \cdots & a_r \end{pmatrix}$$

for some integers $a_1, a_2, \ldots, a_r$ with $a_1 > a_2 > \cdots > a_r \geq 0$. Furthermore, we see that we must have $n = r + 2\sum_{i=1}^{r} a_i$. We then define the map $\lambda = f(\pi) = (2a_1 + 1, 2a_2 + 1, \ldots, 2a_r + 1)$. Clearly, λ is a partition of n into r distinct odd parts. The map f is reversible, and therefore a bijection. $\qquad \square$

1.1.3 Proto-Rogers–Ramanujan Type Identity

We have already seen (Theorem 1.12) that conjugation provides a bijection between partitions of n with l parts and partitions of n with largest part l. A moment's reflection reveals that we may similarly conclude that the number of partitions of n with *at most* l parts equals the number of partitions of n wherein the largest part is *at most* l.

Thus

$$\frac{1}{(1-q)(1-q^2)\cdots(1-q^l)}$$

is the generating function for partitions with at most l parts. We had previously denoted the number of partitions of n into parts at most l as $p_l(n)$. So we now know that $p_l(n)$ also denotes the number of partitions into at most l parts. Since $p_l(n) - p_{l-1}(n)$ counts the number of partitions into *exactly* l parts, the generating function for partitions with exactly l parts is

$$\frac{1}{(1-q)(1-q^2)\cdots(1-q^l)} - \frac{1}{(1-q)(1-q^2)\cdots(1-q^{l-1})}$$
$$= \frac{1-(1-q^l)}{(1-q)(1-q^2)\cdots(1-q^l)} = \frac{q^l}{(1-q)(1-q^2)\cdots(1-q^l)}.$$

In other words, we have just proved

$$\sum_{n=0}^{\infty} p(l,n)q^n = \frac{q^l}{(1-q)(1-q^2)\cdots(1-q^l)}.$$

Since every partition has a well-defined length, the generating function for $p(n)$ must be

$$\sum_{n=0}^{\infty} p(n)q^n = \sum_{l=0}^{\infty} \frac{q^l}{(1-q)(1-q^2)\cdots(1-q^l)}.$$

Taking into account Theorem 1.3, we have proved the identity

$$\sum_{n=0}^{\infty} \frac{q^n}{(1-q)(1-q^2)\cdots(1-q^n)} = \prod_{m=1}^{\infty} \frac{1}{1-q^m}, \tag{1.5}$$

which bears a strong resemblance to (0.1), and thus might be designated a "proto-Rogers–Ramanujan type identity," meaning it is an early prototype of an identity of Rogers–Ramanujan type, discovered by Euler [And76, p. 19, Eq. (2.2.5), with $t = q$] about 150 years before the Rogers–Ramanujan identities, Eqs. (0.1) and (0.2).

1.1.4 Euler's Pentagonal Numbers Theorem

Euler became interested in the reciprocal of the partition generating function, i.e., the infinite product

$$\prod_{m=1}^{\infty}(1-q^m).$$

He expanded this product by direct multiplication and found

$$\prod_{m=1}^{\infty}(1-q^m) = 1 - q - q^2 + q^5 + q^7 - q^{12} - q^{15} + q^{22} + q^{26} - \cdots.$$

It appeared that the coëfficients of the series were all 0, ±1, with nonzero coëfficients at the extended pentagonal numbers. Euler made the following conjecture

Theorem 1.19 (Euler's pentagonal numbers theorem).

$$\prod_{m=1}^{\infty}(1-q^m) = \sum_{j=-\infty}^{\infty} (-1)^j q^{\frac{3}{2}j^2 - \frac{1}{2}j}.$$

Euler found a proof nearly a decade later. We shall not give Euler's proof here, but rather a combinatorial proof based on an observation of Legendre: the infinite product $\prod_{m=1}^{\infty}(1-q^m)$ is *almost* the generating function for partitions with distinct parts, but with a minus sign where the plus should be. Let us multiply out the first few factors to observe the effect of the minus sign:

$$\prod_{m=1}^{\infty}(1-q^m)$$

$$= (1-q^1)\prod_{m=2}^{\infty}(1-q^m)$$

$$= (1-q^1-q^2+q^{2+1})\prod_{m=3}^{\infty}(1-q^m)$$

$$= (1 - q^1 - q^2 + q^{2+1} - q^3 + q^{3+1} + q^{3+2} - q^{3+2+1}) \prod_{m=4}^{\infty} (1 - q^m).$$

So we see that partitions into distinct parts are indeed generated in the exponents, with a coëfficient of 1 whenever the partition has an even number of parts, and with a coëfficient of -1 whenever the number of parts is odd.

Theorem 1.20 (Legendre's partition theoretic interpretation of Euler's pentagonal numbers theorem). *Let $q_e(n)$ denote the number of partitions of n into an even number of parts, all distinct. Let $q_o(n)$ denote the number of partitions of n into an odd number of parts, all distinct. Then*

$$q_e(n) - q_o(n) = \begin{cases} (-1)^j & \textit{if } n \textit{ is of the form } \frac{3}{2}j^2 \pm \frac{1}{2}j \\ 0 & \textit{otherwise} \end{cases}.$$

There is a combinatorial proof of Theorem 1.20, due to Fabian Franklin, a mathematician who earned his Ph.D. under J. J. Sylvester at Johns Hopkins University in 1880.

We will need to define some partition statistics, and a map.

Definition 1.21. For a partition π with l distinct parts given by $(\pi_1, \pi_2, \cdots, \pi_l)$, where $\pi_1 > \pi_2 > \cdots > \pi_l$, let $\rho(\pi)$ denote the length of longest initial run of consecutive integers, i.e. the maximum j such that $\pi_1 = \pi_2 + 1 = \pi_3 + 2 = \cdots = \pi_j + (j-1)$. Let $\sigma(\pi) := \pi_l$, i.e. the smallest part of π. Let

$$f(\pi) = \big(\pi_1 - 1, \pi_2 - 1, \cdots, \pi_{\rho(\pi)} - 1, \pi_{\rho(\pi)+1}, \pi_{\rho(\pi)+2}, \cdots, \pi_l, \rho(\pi)\big).$$

For example, $f\big((6, 5, 3)\big) = (5, 4, 3, 2)$. Graphically, we can think of f as moving the right-most diagonal of dots in the Ferrers diagram to the bottom, as illustrated below.

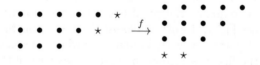

Proof of Theorem 1.20. Let π be a partition of n into l distinct parts. Let us further suppose that $\rho(\pi) \neq l$ and that $\sigma(\pi) \notin \{l, l+1\}$.

Case 1. Suppose $\rho(\pi) < \sigma(\pi)$. Then f maps π to another partition $f(\pi)$ with $l + 1$ distinct parts.

Case 2. Suppose $\rho(\pi) \geq \sigma(\pi)$. Then f^{-1} maps π to another partition $f^{-1}(\pi)$ with $l - 1$ distinct parts. Thus for all partitions of n considered thus far, each "case 1 partition" maps to a "case 2 partition" via f, and each "case 2 partition" maps to a "case 1 partition" via f^{-1}, and the mapping changes the parity of the length of the partition. Thus, of the partitions considered so far, there are the same number of partitions of n into an even number of distinct parts as there are into an odd number of distinct parts.

Now let us consider the partitions we have excluded from the discussion to this point. These are the partitions λ of n into l distinct parts where $\rho(\lambda) = l$ and $\sigma(\lambda) = l$ and the partitions μ of n into l distinct parts where $\rho(\mu) = l$ and $\sigma(\mu) = l + 1$. Thus these partitions are of the form

$$\lambda = (2l, 2l - 1, 2l - 2, \ldots, l + 2, l + 1) \qquad (1.6)$$

and

$$\mu = (2l - 1, 2l - 2, 2l - 3, \ldots, l + 1, l). \qquad (1.7)$$

Notice that even though we have $\rho(\lambda) < \sigma(\lambda)$, $f(\lambda)$ is nonetheless *not* a partition into distinct parts; the two smallest parts of $f(\lambda)$ are the same. Thus partitions of the form λ are "left over" and are not counted in case 1. Notice that

$$n = |\lambda| = (2l) + (2l - 1) + (2l - 2) + \cdots + (l + 2) + (l + 1) = \frac{3}{2}l^2 + \frac{1}{2}l,$$

so this situation only arises when n is of the form $\frac{3}{2}l^2 + \frac{1}{2}l$ and results in $q_e(n) - q_o(n) = (-1)^l$.

Next, $\rho(\mu) = \sigma(\mu)$, but $f^{-1}(\mu)$ is a partition into distinct parts, yet has the same number of parts as μ. Thus partitions of the form μ are "left over" and are not counted in case 2.

$$n = |\lambda| = (2l - 1) + (2l - 2) + (2l - 3) + \cdots + (l + 1) + (l) = \frac{3}{2}l^2 - \frac{1}{2}l.$$

This situation only arises when n is of the form $\frac{3}{2}l^2 - \frac{1}{2}l$ and also results in $q_e(n) - q_o(n) = (-1)^l$. $\qquad \square$

1.1.5 Euler's Recurrence

Theorem 1.22. *For $n > 0$,*

$$p(n) = p(n-1) + p(n-2) - p(n-5) - p(n-7) + p(n-12) + p(n-15) + \cdots$$
$$+ (-1)^{k-1} p\left(n - (k(3k-1)/2)\right) + (-1)^{k-1} p\left(n - (k(3k+1)/2)\right) + \cdots$$
$$(1.8)$$

Proof. The infinite product in Euler's pentagonal numbers theorem and Euler's generating function for $p(n)$ are reciprocals. Consequently, we have

$$\left(\sum_{n=0}^{\infty} p(n)q^n\right)\left(\sum_{j=-\infty}^{\infty} (-1)^j q^{j(3j-1)/2}\right) = 1.$$

Thus,

$$\left(\sum_{n=0}^{\infty} p(n)q^n \right) \left(1 - q - q^2 + q^5 + q^7 - q^{12} - q^{15} + \cdots \right) = 1,$$

or equivalently,

$$\sum_{n=0}^{\infty} p(n)q^n - \sum_{n=0}^{\infty} p(n)q^{n+1} - \sum_{n=0}^{\infty} p(n)q^{n+2} + \sum_{n=0}^{\infty} p(n)q^{n+5} + \sum_{n=0}^{\infty} p(n)q^{n+7} - \cdots = 1,$$

whereupon we shift indices in the sums and recall that $p(n) = 0$ when $n < 0$ to find

$$\sum_{n=0}^{\infty} \left(p(n) - p(n-1) - p(n-2) + p(n-5) + p(n+7) - \cdots \right) q^n = 1.$$

Compare coëfficients of q^n in the extremes and the result follows.   □

Theorem 1.22 provides an efficient algorithm for computing values of $p(n)$. In about 1915, P. A. MacMahon used Theorem 1.22 to calculate $p(n)$ for $0 \le n \le 200$, by hand calculation. It took him about a month, and amazingly the final version did not contain any errors. The table appears as an appendix to Hardy and Ramanujan's paper [HR18], and is credited to MacMahon there. However, in Hardy's "Notice" in Ramanujan's *Collected Papers* [Ram27], Hardy mentions (on p. xxxv) that the table "... was, for the most part, calculated independently by Ramanujan and Major MacMahon; and Major MacMahon was, in general, slightly the quicker and more accurate of the two."

Today, some computer algebra systems use a variant of Theorem 1.22 to compute values of $p(n)$.

1.1.6   Exercises

1. Let $pod(n)$ denote the number of partitions of n where even parts may appear any number of times, but no odd part may appear more than once. Express the generating function of $pod(n)$ as an infinite product.

2. Let $ped(n)$ denote the number of partitions of n where odd parts may appear any number of times, but no even part may appear more than once. Express the generating function of $ped(n)$ as an infinite product.

3. For m a fixed positive integer, let $\rho_m(n)$ denote the number of partitions of n, where no part is a multiple of m. In the literature, such partitions are often called *m-regular partitions*. Express the generating function of $\rho_m(n)$ as an infinite product.

4. Let $t(n)$ denote the number of partitions of n where no part is congruent to 2 (mod 4). Express the generating function of $t(n)$ as an infinite product.

5. For m a fixed positive integer, let $q_m(n)$ denote the number of partitions of m where no part may appear more than $m-1$ times. Express the generating function of $q_m(n)$ as an infinite product.

6. Prove that $p_2(n) = q_2(n)$ for $n \geq 0$. Where have we seen this result before?

7. Prove that $pod(n) = t(n)$ for $n \geq 0$.

8. Prove that $p_3(n) = q_3(n)$ for $n \geq 0$. *Hint:* $1+y+y^2 = \frac{1-y^3}{1-y}$ *when* $y \neq 1$.

9. Prove that $ped(n) = p_4(n) = q_4(n)$ for $n \geq 0$.

10. For m a fixed positive integer, prove that $\rho_m(n) = q_m(n)$ with the aid of the finite geometric series.

11. Prove that the number of partitions of n into distinct parts *not* congruent to 3 (mod 4) equals the number of partitions of n into parts congruent to 1, 5, or 6 (mod 8).

12. Consider the infinite product

$$\prod_{\substack{m \geq 1 \\ m \not\equiv 2 (\text{mod } 4)}} (1-q^m) = \prod_{m=1}^{\infty} (1-q^{2m-1})(1-q^{4m}).$$

(a) Expand the infinite product as a series, and conjecture a result analogous to Euler's pentagonal numbers theorem. What would be an appropriate name for this result?

(b) Prove your conjecture in part (a) by deducing a partition theoretic interpretation analogous to Legendre's interpretation of Euler's pentagonal numbers theorem, and proving it.

1.2 Dedekind's Eta Function and Modular Forms

Let $\mathscr{H}$ denote the upper half of the complex plane

$$\mathscr{H} := \{z \in \mathbb{C} : \Im z > 0\},$$

where $\Im z$ denotes the imaginary part of z. Dedekind's eta function is defined as follows: For $\tau \in \mathscr{H}$,

$$\eta(\tau) := e^{\pi i \tau / 12} \prod_{m=1}^{\infty} (1 - e^{2\pi i \tau m}). \tag{1.9}$$

It is now standard[2] to let $q = e^{2\pi i \tau}$. (We are temporarily thinking of q as a complex variable rather than a formal variable.) Notice that $\tau \in \mathscr{H}$ if and only if $|q| < 1$. Thus we may write

$$\eta(\tau) = q^{1/24} \prod_{m=1}^{\infty} (1 - q^m),$$

and we immediately see that $\eta(\tau)$ differs from the reciprocal of the generating function for $p(n)$ only by an elementary factor. Thus

$$\frac{q^{1/24}}{\eta(\tau)} = \sum_{n=0}^{\infty} p(n) q^n,$$

and, by Euler's pentagonal numbers theorem (Theorem 1.19),

$$\eta(\tau) = \sum_{k=-\infty}^{\infty} (-1)^k q^{(6k-1)^2/24}.$$

The Dedekind eta function is an example of a *modular form*. Barry Mazur once stated in an interview [Lyn97], "Modular forms are functions on the complex plane that are inordinately symmetric. They satisfy so many internal symmetries that their mere existence seem like accidents. But they do exist."

An in-depth introduction to modular forms would take us too far afield of our main line of inquiry, and so we refer the interested reader to the standard references by Fred Diamond and Jerry Shurmann [DS05] or Neil Koblitz [Kob84]. For an introduction to modular forms that emphasizes computation, see the book by L. J. P. Kilford [Kil08] or William A. Stein [Ste07]. For those interested in modular forms primarily in relation to the partition function $p(n)$, the best references include the books by Hans Rademacher [Rad73], Marvin Knopp [Kno93], and Ken Ono [Ono03].

An important property of modular forms is that they satisfy a transformation equation. In particular, $\eta(\tau)$ satisfies the following modular transformation equation [Rad73, p. 163]: Let $a, b, c, d \in \mathbb{Z}$ with $ad - bc = 1$. Then

$$\eta(\tau + b) = e^{\pi i b / 12} \eta(\tau),$$

---

[2]Be careful when consulting the older literature, where sometimes $q = e^{\pi i \tau}$ and $x = e^{2\pi i \tau}$.

and for $c > 0$,

$$\eta\left(\frac{a\tau + b}{c\tau + d}\right) = \epsilon(a, b, c, d)\sqrt{\frac{c\tau + d}{i}}\eta(\tau), \tag{1.10}$$

with

$$\epsilon(a, b, c, d) = \begin{cases} \left(\frac{d}{c}\right)i^{(1-c)/2}\exp\left(\frac{\pi i}{12}(bd(1 - c^2) + c(a + d))\right), & \text{if } 2 \nmid c, \\ \left(\frac{c}{d}\right)\exp\left(\frac{\pi i}{12}(3d + (ac(1 - d^2) + d(b - c)))\right), & \text{if } 2 \nmid d, \end{cases} \tag{1.11}$$

where $\left(\frac{c}{d}\right)$ is the usual Legendre–Jacobi symbol with the convention

$$\left(\frac{c}{d}\right) = \left(\frac{c}{-d}\right)$$

when negative values of d arise.

It turns out that the transformation equation (1.10) is essential to proving a direct formula for $p(n)$, stated later in this book as (2.112).

1.3 Hypergeometric Series

From elementary calculus, we know that a series $\sum_{k=0}^{\infty} t_k$ is called *geometric* if the "term ratio" t_{k+1}/t_k is equal to some constant r for all $k = 0, 1, 2, \ldots$. Furthermore, if $|r| < 1$, then the series converges to $t_0/(1 - r)$, and diverges otherwise. In many ways, geometric series are the most well behaved and useful of all series. After all, we know their convergence behavior and sum (if any), and the proof of the ratio and root tests from elementary calculus rely on comparisons with geometric series. Accordingly, it is a cause for celebration whenever geometric series occur naturally in the course of our work.

Perhaps the next most pleasant type of series after geometric is hypergeometric.

Definition 1.23. A series $\sum_{k=0}^{\infty} t_k$ is called *hypergeometric* if t_{k+1}/t_k is a rational function of k.

Although rarely taught in undergraduate courses these days, most of the commonly used functions in elementary mathematics have representations as hypergeometric series. For example, the Maclaurin series representation of the exponential function is

$$e^z = \sum_{k=0}^{\infty} \frac{z^k}{k!}, \tag{1.12}$$

which has term ratio $z/(k + 1)$, and is thus hypergeometric.

Since the term ratio t_{k+1}/t_k of a hypergeometric series $\sum_{k=0}^{\infty} t_k$ is a rational function of k, it can, without loss of generality, be written as

$$\frac{t_{k+1}}{t_k} = \frac{(k+a_1)(k+a_2)\cdots(k+a_r)}{(k+1)(k+b_1)(k+b_2)\cdots(k+b_s)} z, \qquad (1.13)$$

since any polynomial can be factored completely into linear factors over the complex numbers.

Remark 1.24. The factor $(k+1)$ in the denominator of the right hand side of (1.13) is included for historical reasons, i.e. most hypergeometric series of interest have a term ratio which includes the factor $(k+1)$ in the denominator. If the polynomial in the denominator of the term ratio of a given hypergeometric series does *not* contain the factor $(k+1)$, then we compensate for this by setting one of the a_i to 1.

Consider the hypergeometric series $\sum_{k\geq 0} t_k$ such that $t_0 = 1$, and

$$\frac{t_{k+1}}{t_k} = \frac{(k+a)(k+b)}{(k+1)(k+c)} z. \qquad (1.14)$$

Notice that we can iterate (1.14) to find that

$$\sum_{k\geq 0} t_k = 1 + \frac{ab}{c}z + \frac{a(a+1)b(b+1)}{2c(c+1)}z^2 + \frac{a(a+1)(a+2)b(b+1)(b+2)}{3!c(c+1)(c+2)}z^3 + \cdots .$$

$$(1.15)$$

Products of the form $a(a+1)(a+2)\cdots$ occur so frequently in the study of hypergeometric series that they are given a compact notation.

Definition 1.25. The *rising factorial* or *Pochhammer symbol* $(a)_k$ is given by

$$(a)_k := \begin{cases} \prod_{j=0}^{k-1}(a+j) & \text{if } k \text{ is a positive integer,} \\ 1 & \text{if } k = 0, \\ \prod_{j=1}^{k}(a-j)^{-1} & \text{if } k \text{ is a negative integer.} \end{cases}$$

Note that $(1)_k = k!$.

Remark 1.26. In analysis, it is standard to define

$$\Gamma(z) := \int_0^{\infty} t^{z-1}e^{-t}\, dt,$$

for $\Re z > 0$, and to extend the domain of definition to $\mathbb{C} \setminus \{0, -1, -2, -3, \dots\}$. One can show using integration by parts that the functional equation

$$\Gamma(z+1) = z\Gamma(z)$$

is satisfied, and the initial condition $\Gamma(1) = 1$ may be verified by direct computation. Thus for positive integers n, $\Gamma(n) = (n-1)!$. For our purposes, we will use the symbol $\Gamma(n+1)$ interchangeably with $n!$ or $(1)_n$.

The hypergeometric series with initial ($k = 0$ term) of 1, and with term ratio given by (1.13), is

$$
_rF_s\left[\begin{matrix} a_1, a_2, \ldots, a_r \\ b_1, b_2, \ldots, b_s \end{matrix}; z\right] := \sum_{k=0}^{\infty} \frac{(a_1)_k (a_2)_k \cdots (a_r)_k}{k!(b_1)_k (b_2)_k \cdots (b_s)_k} z^k. \tag{1.16}
$$

The notation on the left hand side of (1.16) is the standard notation for hypergeometric series, and is sometimes referred to as the *normal form*. The $a_1, a_2, \ldots, a_r$ are called the *numerator parameters*; the $b_1, \ldots, b_s$ are the *denominator parameters*; and z is called the *argument* of the hypergeometric series.

Example 1.27. Show that the Maclaurin series for $\sin z$ is hypergeometric, and find its normal form.
Solution. From elementary calculus, we know that

$$
\sin z = \sum_{k=0}^{\infty} \frac{(-1)^k z^{2k+1}}{(2k+1)!}.
$$

Thus, the term ratio is

$$
\frac{(-1)^{k+1} z^{2k+3}}{(2k+3)!} \cdot \frac{(2k+1)!}{(-1)^k z^{2k+1}} = -\frac{z^2}{(2k+3)(2k+2)}
$$
$$
= \frac{1}{(k+1)(k+\frac{3}{2})}\left(-\frac{z^2}{4}\right),
$$

which is a rational function of k. Also, note that the $k = 0$ term of the series is not 1, but rather it is z. Thus the normal form of the series is

$$
z \, _0F_1\left[\begin{matrix} - \\ \frac{3}{2} \end{matrix}; -\frac{z^2}{4}\right].
$$

Some additional examples are:

$$
\frac{1}{1-z} = \sum_{k=0}^{\infty} z^k = \, _1F_0\left[\begin{matrix} 1 \\ - \end{matrix}; z\right], \tag{1.17}
$$

$$
\frac{1}{(1-z)^\alpha} = \sum_{k=0}^{\infty} \frac{(\alpha)_k}{k!} z^k = \, _1F_0\left[\begin{matrix} \alpha \\ - \end{matrix}; z\right], \tag{1.18}
$$

$$
e^z = \sum_{k=0}^{\infty} \frac{z^k}{k!} = \, _0F_0\left[\begin{matrix} - \\ - \end{matrix}; z\right], \tag{1.19}
$$

$$
\arctan z = \sum_{k=0}^{\infty} \frac{(-1)^k}{2k+1} z^{2k+1} = z \, _2F_1\left[\begin{matrix} \frac{1}{2}, 1 \\ \frac{3}{2} \end{matrix}; -z^2\right]. \tag{1.20}
$$

1.3.1 Gauss' Hypergeometric Series

The hypergeometric series

$$
{}_2F_1\left[\begin{matrix}a, b\\c\end{matrix}; z\right]
\tag{1.21}
$$

was studied extensively by C. F. Gauss, and he delivered a famous lecture on such series in January, 1812. In fact, until fairly recently, the term "hypergeometric series" *only* referred to the ${}_2F_1$, while the ${}_rF_s$ for $r > 2$ or $s > 1$ was called a "generalized hypergeometric series."

Theorem 1.28 (Gauss' Hypergeometric Summation Formula).

$$
{}_2F_1\left[\begin{matrix}a, b\\c\end{matrix}; 1\right] = \frac{\Gamma(c)\Gamma(c - a - b)}{\Gamma(c - a)\Gamma(c - b)}.
$$

We will not prove Theorem 1.28 here, as it can be proved easily using the techniques introduced in the next section.

For additional information on Gauss' hypergeometric series, including its relation to differential equations, see e.g. [Bai35] or [Sla66].

1.3.2 Classical Hypergeometric Summation Formulæ

Besides Gauss' sum, there are many known hypergeometric summation formulæ. We list just a few examples below. A uniform method of proof, due to Wilf and Zeilberger, for these and indeed all hypergeometric identities will be provided in the subsequent sections. Let n denote a positive integer throughout.

Theorem 1.29 (Chu–Vandermonde).

$$
{}_2F_1\left[\begin{matrix}-n, b\\c\end{matrix}; 1\right] = \frac{(c - b)_n}{(c)_n}.
$$

Historical Note 1.30. Theorem 1.29 was discovered in 1770 by the French mathematician Alexandre-Théophile Vandermonde and was called "Vandermonde's sum" in the literature for many years. Recently, however, it was noticed that this identity had been discovered more than four and a half centuries earlier and had appeared in a book written in 1303 by the Chinese mathematician Chu Shih-Chieh, so we now call it the "Chu–Vandermonde sum." This same book from 1303 contains an illustration of what we call "Pascal's triangle" and refers to it as an "ancient method."

Theorem 1.31 (Kummer).

$$
{}_2F_1\left[\begin{matrix}1 - c - 2n, -2n\\c\end{matrix}; -1\right] = (-1)^n \frac{(2n)!(c - 1)!}{n!(c + n - 1)!}
$$

Theorem 1.32 (Pfaff–Saalschütz).

$$_3F_2\left[\begin{matrix} a, b, -n \\ c, 1+a+b-c-n \end{matrix}; 1\right] = \frac{(c-a)_n(c-b)_n}{(c)_n(c-a-b)_n}$$

Historical Note 1.33. The earliest known use of the term "hypergeometric" is in a 1655 book by Oxford professor John Wallis [Wal55]. Wallis applied the term to the series

$$1 + x + x(x+1) + x(x+1)(x+2) + \cdots.$$

Gauss inaugurated the systematic study of the series (1.15) in his famous address to the Royal Society in Göttingen on January 20, 1812 [Gau13]. Gauss used the notation $F[a, b; c; z]$ for what we would write as $_2F_1\left[\begin{matrix} a,b \\ c \end{matrix}; 1\right]$.

The standard (and highly recommended) twentieth century works on hypergeometric series were published by W. N. Bailey in 1935 [Bai35] and L. J. Slater in 1966 [Sla66]. These works treat hypergeometric series from the analytic viewpoint.

1.3.3 Exercises

Show that the following functions can be represented by hypergeometric series, and find their normal forms:

1. $\cos z$

2. $\arcsin z$

3. $\log(1-z)$

1.4 Wilf–Zeilberger Theory of Hypergeometric Summation

1.4.1 Zeilberger's Creative Telescoping Algorithm

Although hypergeometric series were first studied systematically by Gauss, one of the most significant advances in the study of hypergeometric series occurred quite recently. Doron Zeilberger designed an algorithm by which identities involving hypergeometric series can be proved algorithmically. Consider a series

$$\sum_{k\geq 0} F(n, k) \tag{1.22}$$

where $F(n,k)$ is a summand which is *hypergeometric in both n and k*, that is, $F(n+1,k)/F(n,k)$ and $F(n,k+1)/F(n,k)$ are both rational functions of n and k, where $F(n,k)=0$ whenever $k<0$ or $k>M$ for some integer M. Think of both the left hand side and right hand side of (1.22) as a sequence of rational functions indexed by n. If we can show that both sides of (1.22) satisfy the same recurrence and initial conditions, then we will have proved their equality.

If we could find some nice function $G(n,k)$ where $\lim_{k\to\pm\infty} G(n,k)=0$ and coëfficients $a_0(n)$ and $a_1(n)$ (independent of k) such that

$$a_0(n)F(n,k)+a_1(n)F(n+1,k)=G(n,k)-G(n,k+1), \qquad (1.23)$$

then by summing (1.23) over all k, we would find

$$a_0(n)f(n+1)+a_1(n)f(n)=0, \qquad (1.24)$$

where $f(n)=\sum_k F(n,k)$, since the right hand side telescopes to zero.

If no such $a_0(n)$, $a_1(n)$, and $G(n,k)$ exist, we would then look for $a_0(n)$, $a_1(n)$, $a_2(n)$, and $G(n,k)$ such that

$$a_0(n)F(n,k)+a_1(n)F(n+1,k)+a_2(n)F(n+2,k)=G(n,k+1)-G(n,k).$$
$$(1.25)$$

Zeilberger's algorithm guarantees that there exists a positive integer J so that

$$\sum_{j=0}^{J} a_j(n)F(n+j,k)=G(n,k+1)-G(n,k) \qquad (1.26)$$

for some $a_0(n),\ldots,a_J(n)$ and $G(n,k)$. Zeilberger and Moa Apagodo found a sharp upper bound for J [MZ05]. Furthermore, it turns out that $G(n,k)=F(n,k)R(n,k)$ for some rational function $R(n,k)$.

Notice that by dividing both sides of (1.26) by $F(n,k)$ we obtain the equivalent equation

$$\sum_{j=0}^{J} a_j(n)\frac{F(n+j,k)}{F(n,k)}=\frac{F(n,k+1)R(n,k+1)}{F(n,k)}-R(n,k), \qquad (1.27)$$

which is an easily checked identity of rational functions.

The rational function $R(n,k)$ is called the *certificate* for the recurrence. For if one has in hand the recurrence (i.e. $a_0(n),a_1(n),\ldots,a_J(n)$) and the certificate $R(n,k)$, then (1.27) follows, which is equivalent to (1.26).

Formally, we record the procedure as follows.

Algorithm 1.34 (Zeilberger's creative telescoping algorithm).

INPUT: A summand $F(n,k)$ with finite support which is hypergeometric in both n and k.

OUTPUT: Expressions $a_0, a_1, \ldots, a_J$ independent of k and a rational function $R(n, k)$, such that

$$\sum_{j=0}^{J} a_j F(n+j, k) = R(n, k+1)F(n, k+1) - R(n, k)F(n, k)$$

for some J.

Example 1.35. Let

$$F(n, k) = \frac{(n-k+1)_k}{k!} x^k y^{n-k}.$$

Zeilberger's algorithm finds a first order recurrence with $a_0 = x+y$, $a_1 = -1$, certified by

$$R(n, k) = \frac{ky}{n-k+1}.$$

The full proof is thus

$$(x+y) - \frac{(n+1)y}{n-k+1} = \frac{(n-k)x(k+1)y}{(k+1)y(n-(k+1)+1)} - \frac{ky}{n-k+1}$$

$$\implies (x+y)F(n, k) - F(n+1, k) = G(n, k+1) - G(n, k)$$

$$\implies (x+y)\sum_k F(n, k) - \sum_k F(n+1, k) = 0$$

$$\implies (x+y)\sum_k F(n, k) = \sum_k F(n+1, k). \tag{1.28}$$

Furthermore, Eq. (1.28) together with the observation that $\sum_k F(0, k) = 1$ allows us to conclude that

$$\sum_k F(n, k) = (x+y)^n,$$

and thus we have proved the binomial theorem.

Theorem 1.36 (Binomial theorem). *For n a nonnegative integer,*

$$\sum_{k=0}^{n} \binom{n}{k} x^k y^{n-k} = (x+y)^n, \tag{1.29}$$

where

$$\binom{n}{k} = \begin{cases} \dfrac{(n-k+1)_k}{k!} = \dfrac{n!}{(n-k)!k!} & \text{if } 0 \leq k \leq n, \\ 0 & \text{otherwise.} \end{cases}$$

Next, we prove another important hypergeometric summation formula.

Theorem 1.37 (Pfaff–Saalschütz sum).

$$_3F_2\left[\begin{matrix} a,b,-n \\ c,1+a+b-n-c \end{matrix};1\right] = \frac{(c-a)_n(c-b)_n}{(c)_n(c-a-b)_n} \qquad (1.30)$$

Proof. Let $f(n)$ denote the left hand side of (1.30). Zeilberger's algorithm reveals that $f(n)$ satisfies the recurrence

$$f(n+1) = \frac{(c-a+n)(c-b+n)}{(c+n)(c-a-b+n)}f(n), \qquad (1.31)$$

certified by

$$R(n,k) = \frac{k(c+k-1)(a+b-c+k-n)}{n-k+1}.$$

Clearly, the right hand side of (1.30) also satisfies (1.31). Finally, both sides of (1.30) equal 1 when $n=0$, thus (1.30) follows. □

Corollary 1.38 (The Chu-Vandermonde sum).

$$_2F_1\left[\begin{matrix} a,-n \\ c \end{matrix};1\right] = \frac{(c-a)_n}{(c)_n} \qquad (1.32)$$

Proof. Let $b \to \infty$ in (1.30). □

1.4.2   The WZ Method

Given only a hypergeometric summand $F(n,k)$ with finite support, Zeilberger's algorithm finds a recurrence satisfied by that summand and a rational function proof certificate. If it turns out that the recurrence found is first order, this can in turn be iterated to give a finite product expression for the sum of $F(n,k)$ over k. In many circumstances, both sides of the identity (the hypergeometric series and its closed form sum) are known in advance (at least conjecturally). We now turn our attention to an algorithm, due to Herbert Wilf and Doron Zeilberger, and named in honor of two famous complex variables, which allows a computer to generate proofs of hypergeometric summation formulæ. [WZ90], [PWZ96].

Let $U(n,k)$ be a hypergeometric term, i.e., suppose that the ratios

$$U(n+1,k)/U(n,k)$$

and

$$U(n,k+1)/U(n,k)$$

are both rational functions of n and k, and we want to prove that

$$\sum_{k\in\mathbb{Z}} U(n,k) = f(n), \qquad (1.33)$$

where $f(n)$ is a nonzero expression that is hypergeometric in n.

Divide (1.33) through by $f(n)$ to obtain the equivalent equation[3]

$$\sum_{k\in\mathbb{Z}} F(n,k) = 1, \tag{1.34}$$

where

$$F(n,k) := \frac{U(n,k)}{f(n)}.$$

Now notice that the right hand side of (1.34) is independent of n, so it is equally true that

$$\sum_{k\in\mathbb{Z}} F(n+1,k) = 1. \tag{1.35}$$

Subtracting (1.34) from (1.35), we have

$$\sum_{k\in\mathbb{Z}} \Big(F(n+1,k) - F(n,k) \Big) = 0. \tag{1.36}$$

It would be great if there were a function $G(n,k)$ such that

$$F(n+1,k) - F(n,k) = G(n,k+1) - G(n,k) \tag{1.37}$$

and

$$\lim_{k\to\pm\infty} G(n,k) = 0.$$

If such a $G(n,k)$ existed, then we would have

$$\sum_{k=-L}^{K} \Big(F(n+1,k) - F(n,k) \Big) = \sum_{k=-L}^{K} \Big(G(n,k+1) - G(n,k) \Big)$$
$$= G(n,K+1) - G(n,-L),$$

and upon letting $K, L \to \infty$, we would immediately obtain (1.36).

As it turns out, under very mild hypotheses, such a $G(n,k)$ will exist and can be determined algorithmically using just $F(n,k)$ as input and, furthermore, $G(n,k)$ will turn out to be a rational function $R(n,k)$ multiple of $F(n,k)$. Thus with $G(n,k) = F(n,k)R(n,k)$, we may divide (1.37) by $F(n,k)$ to obtain the equivalent equation

$$\frac{F(n+1,k)}{F(n,k)} - 1 = \frac{F(n,k+1)R(n,k+1)}{F(n,k)} - R(n,k), \tag{1.38}$$

which simply asserts the equality of two rational functions. The function $R(n,k)$ is called the *WZ certificate*. The WZ algorithm has been implemented

---

[3]Professor Zeilberger refers to this as the "Wilf trick." But the reader should keep in mind that the definition of a *method* is a trick that has worked more than once.

in Maple and Mathematica. Starting with version 8 of Maple, Zeilberger's algorithm and the WZ theory are built into the SumTools package. Also see Zeilberger's home page for additional WZ-theory and q-WZ-theory Maple packages. Mathematica implementations are available to researchers and students free of charge from the Research Institute for Symbolic Computation at Johannes Kepler Univesrity in Linz, Austria. Search the web for "RISC algorithmic combinatorics software."

To see the WZ method in action, let us use the method to prove the well-known identity

$$\sum_{k=0}^{n} \binom{n}{k} = 2^n, \tag{1.39}$$

for n a nonnegative integer.

WZ proof. $F(n,k) = \binom{n}{k}2^{-n}$. Let $R(n,k) = \frac{-k}{2(n-k+1)}$. Substitution of $F(n,k)$ and $R(n,k)$ into (1.38) gives

$$\frac{n+1}{2(n-k+1)} - 1 = \frac{-kn+k^2-n+k}{(k+1)(2n-2k)} + \frac{k}{2n+2-2k}, \tag{1.40}$$

which simplifies further to the tautology

$$-\frac{n+1-2k}{2(n-k+1)} = -\frac{n+1-2k}{2(n-k+1)}.$$

Thus we have the following chain of reasoning:

$$-\frac{n+1-2k}{2(n-k+1)} = -\frac{n+1-2k}{2(n-k+1)}$$

$$\implies \frac{n+1}{2(n-k+1)} - 1 = \frac{-kn+k^2-n+k}{(k+1)(2n-2k)} + \frac{k}{2n+2-2k}$$

$$\implies \frac{\binom{n+1}{k}2^{-(n+1)}}{\binom{n}{k}2^{-n}} - 1 = \frac{-kn+k^2-n-k}{(k+1)(2n-2k)} + \frac{k}{2n+2-2k}$$

$$\implies \binom{n+1}{k}2^{(n+1)} - \binom{n}{k}2^{-n} = \binom{n}{k+1}2^{-n}\frac{-(k+1)}{2(n-k)} - \frac{k}{2(n-k+1)}\binom{n}{k}2^{-n}$$

$$\implies F(n+1,k) - F(n,k) = F(n,k+1)R(n,k+1) - F(n,k)R(n,k)$$

$$\implies F(n+1,k) - F(n,k) = G(n,k+1) - G(n,k)$$

$$\implies \sum_{k\in\mathbb{Z}}(F(n+1,k) - F(n,k)) = 0$$

$$\implies \sum_{k\in\mathbb{Z}}F(n+1,k) = \sum_{k\in\mathbb{Z}}F(n,k)$$

$$\implies \sum_{k\in\mathbb{Z}}F(n,k) \text{ is constant for all } n \geq 0$$

$$\implies \sum_{k \in \mathbb{Z}} F(n,k) = \sum_{k \in \mathbb{Z}} F(0,k) = \sum_{k \in \mathbb{Z}} \binom{0}{k} = 1$$

$$\implies \sum_{k \in \mathbb{Z}} \binom{n}{k} 2^{-n} = 1$$

$$\implies \sum_{k \in \mathbb{Z}} \binom{n}{k} = 2^n,$$

and since $\binom{n}{k} = 0$ when $k < 0$ or $k > n$, we have (1.39). □

Earlier it was noted that the WZ certificate, which is the key that unlocks the above proof, can be determined algorithmically and indeed the algorithm is explained in [PWZ96, §6.3]. However some readers might be uncomfortable with quantities that "magically" appear on the computer screen since a given computer algebra system and the associated software may contain bugs. This really should not be a concern, since once the correct $R(n,k)$ is produced (by *any* method), its correctness will be guaranteed by verifying the associated algebraic identity.

Let us examine a method by which we can *guess* $R(n,k)$ (or equivalently $G(n,k)$) without a computer.

Notice that $G(n,k) = 0$ when $k < 0$, since $G(n,k) = F(n,k)R(n,k)$ and $F(n,k) = \binom{n}{k}2^{-n} = 0$ when $k < 0$. Next, solve (1.37) for $G(n,k+1)$:

$$G(n,k+1) = F(n+1,k) - F(n,k) + G(n,k). \tag{1.41}$$

Setting $k = -1$ in (1.41), we find

$$G(n,0) = F(n+1,-1) - F(n,-1) + G(n,-1) = 0 - 0 + 0 = 0.$$

Next, set $k = 0$ in (1.41), to find

$$G(n,1) = F(n+1,0) - F(n,0) + G(n,0) = \frac{1}{2^{n+1}} - \frac{1}{2^n} + 0 = -\frac{1}{2^{n+1}}.$$

Next, set $k = 1$ in (1.41), to find

$$G(n,2) = F(n+1,1) - F(n,1) + G(n,1) = \frac{\binom{n+1}{1}}{2^{n+1}} - \frac{\binom{n}{1}}{2^n} - \frac{1}{2^{n+1}} = \frac{-n}{2^{n+1}}.$$

Next, set $k = 2$ in (1.41), to find

$$G(n,3) = F(n+1,2) - F(n,2) + G(n,2) = \frac{\binom{n+1}{2}}{2^{n+1}} - \frac{\binom{n}{2}}{2^n} = \frac{-\binom{n}{2}}{2^{n+1}}.$$

Iterate to find

$$G(n,4) = -\frac{\binom{n}{3}}{2^{n+1}},$$

$$G(n, 5) = -\frac{\binom{n}{4}}{2^{n+1}}, \text{ etc.}$$

and continue until you are confident that

$$G(n, k) = -\frac{\binom{n}{k-1}}{2^{n+1}},$$

which implies that

$$R(n, k) = -\frac{k}{2(n - k + 1)}.$$

1.4.3   WZ Certificates for Classical Hypergeometric Summation Formulæ

Remark 1.39. Zeilberger's algorithm produces a recurrence and rational function for each side of the identity, while in contrast the "WZ" method effectively combines both sides of the equation into one, and thus only requires a single certificate function.

Theorem 1.29 is WZ-certified by

$$R(n, k) = \frac{k(c + k - 1)}{(b - n - c)(n - k + 1)}.$$

Theorem 1.31 is WZ-certified by

$$R(n, k) = \frac{(k + 1)(k + c)}{n(n + 1 - c)}.$$

Theorem 1.32 is WZ-certified by

$$R(n, k) = \frac{k(c + k - 1)(a + b - c + k - n)}{(b - c - n)(a - c - n)(n - k + 1)}.$$

1.4.4   Shift Operators

Define the *n-forward shift operator* η such that

$$\eta F(n, k) = F(n + 1, k),$$

thus

$$\eta^j F(n, k) = F(n + j, k)$$

for any $j \in \mathbb{Z}$.

Further define the *k-forward shift operator* κ such that

$$\kappa F(n, k) = F(n, k + 1).$$

Thus the WZ equation (1.37) may be written

$$\eta F(n,k) = \kappa G(n,k).$$

The *Fibonacci sequence* $\{F_n\}_{n=0}^{\infty}$ is defined by $F_0 = 0$, $F_1 = 1$ and

$$F_{n+1} = F_n + F_{n-1} \tag{1.42}$$

for $n > 0$. The latter may be written

$$(\eta - 1 - \eta^{-1})F_n = 0,$$

(where $1 = \eta^0$ is the identity operator) illustrating that a linear recurrence operator in n may be viewed as a Laurent polynomial in η. Further, a recurrence operator $A(\eta, \kappa)$ is an *annihilating operator* of $F(n,k)$ if

$$A(\eta, \kappa)F(n,k) = 0.$$

The Fibonacci sequence is also annihilated by the operator $\eta^2 - \eta - 1$, since we may shift the index in (1.42) and write

$$F_{n+2} - F_{n+1} - F_n = 0.$$

The Fibonacci sequence satisfies a recurrence of order 2, and no recurrence of order less than 2, but one can easily find recurrences of order $3, 4, 5$, etc. satisfied by the Fibonacci sequence using the algebra of operators.

Since

$$(\eta^2 - \eta - 1)F_n = 0,$$

it must also be the case that

$$B(\eta)(\eta^2 - \eta - 1)F_n = B(\eta)0 = 0$$

for operators $B(\eta)$. So, to produce a third order recurrence satisfied by F_n, simply take $B(\eta)$ to be a first degree polynomial in η, say, $B(\eta) = \eta + 2$. Then

$$\begin{aligned}
0 &= B(\eta)0 \\
&= B(\eta)(\eta^2 - \eta - 1)F_n \\
&= (\eta + 2)(\eta^2 - \eta - 1)F_n \\
&= (\eta^3 + \eta^2 - 3\eta - 2)F_n,
\end{aligned}$$

so we find that the Fibonacci sequence satisfies the third order recurrence

$$F_{n+3} = -F_{n+2} + 3F_{n+1} + 2F_n.$$

Thus, composition of operators may be achieved via multiplication of the corresponding Laurent polynomials in η. Be careful when multiplying operators to remember that this multiplication is, in general, *not* commutative. The idea of "multiplying" and "factoring" operators will play an important role later in our studies. For an enlightening discussion on the algebra of shift operators, see the book $A = B$ by Petkovšek, Wilf, and Zeilberger [PWZ96, Chapter 9].

1.5 q-Analogs

1.5.1   Natural q-Analogs

What is meant by a q-analog? Strictly speaking, the only requirement for $A(q)$ to be a q-analog of α is that

$$\lim_{q \to 1} A(q) = \alpha. \tag{1.43}$$

Using this criterion alone, we can say that $\sqrt{q} + 3q^2 + 5$ is a q-analog of the number 9, which is true, but one would be very hard pressed to find a situation where $\sqrt{q} + 3q^2 + 5$ is a "natural" or "useful" q-analog of 9. However, we do see q-analogs arise naturally in some rather familiar settings.

One such "natural" q-analog of the number α, introduced by Eduard Heine [Hei78], is

$$[\alpha]_q := \frac{1 - q^\alpha}{1 - q}. \tag{1.44}$$

Why is (1.44) a natural q-analog of α? Among other reasons, it plays the role of the number α in a straightforward generalization of a number of familiar settings. For example, consider the derivative a single variable function $f(x)$ from elementary calculus.

We know that

$$
\begin{aligned}
Df(x) &= \lim_{h \to 0} \frac{f(x+h) - f(x)}{h} \\
&= \lim_{y \to x} \frac{f(y) - f(x)}{y - x} \quad \text{(by letting } y = x + h) \\
&= \lim_{qx \to x} \frac{f(qx) - f(x)}{qx - x} \quad \text{(by letting } y = qx) \\
&= \lim_{q \to 1} \frac{f(x) - f(qx)}{x(1 - q)}.
\end{aligned}
$$

Let us now define the *q-derivative operator* as

$$D_q f(x) := \frac{f(x) - f(qx)}{x(1 - q)}. \tag{1.45}$$

Certainly, D_q is a q-analog of the derivative operator since it satisfies (1.43). Notice that

$$D_q x^\alpha = \frac{x^\alpha - (xq)^\alpha}{x(1 - q)} = \left(\frac{1 - q^\alpha}{1 - q} \right) \frac{x^\alpha}{x} = [\alpha]_q x^{\alpha - 1}.$$

Thus, $[\alpha]_q = \frac{1 - q^\alpha}{1 - q}$ plays the same role in the "q-power rule" as α does in the ordinary power rule from elementary calculus.

For $k \in \mathbb{Z}_+$, let $[k]_q! := [k]_q[k-1]_q \cdots [2]_q[1]_q$, and let $[0]_q! := 1$.
Now let us define a q-analog of the exponential function as

$$e_q(z) := \sum_{k=0}^{\infty} \frac{z^k}{[k]_q!}. \tag{1.46}$$

We get the following desirable result.

Proposition 1.40.

$$D_q e_q(z) = e_q(z) \tag{1.47}$$

Proof. Exercise. ☐

Of course (1.46) is not the only possible q-exponential function. Another
is

$$E_q(z) := \sum_{k=0}^{\infty} \frac{z^k q^{k(k-1)/2}}{[k]_q!}. \tag{1.48}$$

1.5.2 Exercises

1. Prove Proposition 1.40.

2. Show that $E_q(z)$ is, in fact, a q-analog of e^z.

3. Show that although $D_q E_q(z) \neq E_q(z)$, we nonetheless have

$$D_q E_q(z) = E_q(zq),$$

which is a q-analog of the formula $De^z = e^z$.

4. Show that $e_{1/q}(z) = E_q(x)$.

1.6 Basic Hypergeometric Series

Heine generalized Gauss' hypergeometric series (the $_2F_1$) by introducing
the *basic analog (or q-analog) of the complex number* α:

$$[\alpha]_q := \frac{1 - q^\alpha}{1 - q}.$$

Thus a basic analog of the rising factorial

$$(\alpha)_k = \alpha(\alpha + 1)(\alpha + 2) \cdots (\alpha + k - 1)$$

is given by

$$([\alpha]_q)_k = [\alpha]_q[\alpha+1]_q\cdots[\alpha+k-1]_q = \frac{(1-q^\alpha)(1-q^{\alpha+1})\cdots(1-q^{\alpha+k-1})}{(1-q)^k}.$$

In practice, however, rising factorials often appear in ratios in identical quantities in the numerator and denominator, so the $(1-q)$ factors cancel.

Also, following G. N. Watson, q^α is normally replaced by a.

Definition 1.41. A series $\sum_{k\geq 0} u_{n,k,q}$ is *basic hypergeometric (with base q),* or *q-hypergeometric,* if

$$\frac{u_{n+1,k,q}}{u_{n,k,q}}$$

and

$$\frac{u_{n,k+1,q}}{u_{n,k,q}}$$

are both rational functions of q^n and q^k.

Definition 1.42. Let $k \in \mathbb{Z}$. We define

$$(a;q)_k := \begin{cases} (1-a)(1-aq)(1-aq^2)\cdots(1-aq^{k-1}) & \text{if } k > 0 \\ 1 & \text{if } k = 0 \\ \frac{1}{(1-aq^{-1})(1-aq^{-2})\cdots(1-aq^k)} & \text{if } k < 0 \end{cases}.$$

Remark 1.43. The symbol $(a;q)_k$ is called a *q-Pochhammer symbol,* which denotes a "rising q-factorial" or "q-shifted factorial."

Definition 1.44. The *basic hypergeometric series*

$$_r\phi_s \begin{bmatrix} a_1, a_2, \ldots, a_r \\ b_1, b_2, \ldots, b_s \end{bmatrix} ; q, z \end{bmatrix}$$

$$:= \sum_{k=0}^{\infty} \frac{(a_1;q)_k(a_2;q)_k\cdots(a_r;q)_k}{(q;q)_k(b_1;q)_k\cdots(b_s;q)_k} z^k \left((-1)^k q^{k(k-1)/2}\right)^{1+s-r}.$$

The basic hypergeometric series

$$_{r+1}\phi_r \begin{bmatrix} a_1, a_2, \ldots, a_{r+1} \\ b_1, b_2, \ldots, b_r \end{bmatrix} ; q, z \end{bmatrix}$$

is called *well-poised* if

$$qa_1 = a_2b_1 = a_3b_2 = \cdots = a_{r+1}b_r,$$

and *very-well-poised* if additionally $a_2 = q\sqrt{a_1}$ and $a_3 = -q\sqrt{a_1}$. It is called *balanced* (or *Saalschützian*) if $b_1b_2\cdots b_r = qa_1a_2\cdots a_{r+1}$.

Remark 1.45. Let $a_i = q^{\alpha_i}$ for $i = 1, 2, \ldots, r$ and $b_j = q^{\beta_j}$ for $j = 1, 2, \ldots, s$. Then

$$\lim_{q \to 1^-} {}_r\phi_s \left[\begin{matrix} a_1, a_2, \ldots, a_r \\ b_1, b_2, \ldots, b_s \end{matrix} ; q, z \right] = {}_rF_s \left[\begin{matrix} \alpha_1, \alpha_2, \ldots, \alpha_r \\ \beta_1, \beta_2, \ldots, \beta_s \end{matrix} ; z \right].$$

Many of the classical hypergeometric summation formulæ, including those in the previous section, have q-analogs. The WZ theory translates almost word-for-word to the q-case, with the roles of n and k now played by q^n and q^k, respectively. Alternatively, the roles of n and k could be played by $[n]_q$ and $[k]_q$.

Theorem 1.46 (q-Chu–Vandermonde).

$$_2\phi_1 \left[\begin{matrix} a, q^{-n} \\ c \end{matrix} ; q, \frac{cq^n}{a} \right] = \frac{(c/a; q)_n}{(c; q)_n}.$$

Proof. The q-WZ certificate is

$$R_q(n, k) = \frac{cq^{n-1}(1 - aq^k)(1 - q^{n-k})}{a(1 - q^n)(1 - cq^{n-1})}.$$

$\square$

We record the next two results here because they are q-analogs of well-known hypergeometric summations. Their proofs, however, rely on two results that will be proved later.

Theorem 1.47 (Heine's q-Gauss sum).

$$_2\phi_1 \left[\begin{matrix} a, b \\ c \end{matrix} ; q, \frac{c}{ab} \right] = \frac{(c/a; q)_\infty (c/b; q)_\infty}{(c; q)_\infty (c/ab; q)_\infty}.$$

Proof.

$$\begin{aligned}
_2\phi_1 \left[\begin{matrix} a, b \\ c \end{matrix} ; q, c/ab \right] &= \frac{(b; q)_\infty (c/b; q)_\infty}{(c; q)_\infty (c/ab; q)_\infty} {}_1\phi_0 \left[\begin{matrix} c/ab \\ - \end{matrix} ; q, b \right] \\
&= \frac{(c/a; q)_\infty (c/b; q)_\infty}{(c; q)_\infty (c/ab; q)_\infty} \text{ (by Thm. 1.64).}
\end{aligned}$$

$\square$

Theorem 1.48 (q-Kummer sum; due to Bailey [Bai41] and Daum [Dau42]).

$$_2\phi_1 \left[\begin{matrix} a, b \\ aq/b \end{matrix} ; q, -\frac{q}{b} \right] = \frac{(-q; q)_\infty (aq; q^2)_\infty (aq^2/b^2; q^2)_\infty}{(-q/b; q)_\infty (aq/b; q)_\infty}.$$

Proof.

$$_2\phi_1\left[\begin{matrix}a,b\\aq/b\end{matrix};q,-q/b\right]=\frac{(a;q)_\infty(-q;q)_\infty}{(aq/b;q)_\infty(-q/b;q)_\infty}{}_2\phi_1\left[\begin{matrix}q/b,-q/b\\-q\end{matrix};q,a\right]$$

$$\text{(by Eq. (1.93))}$$

$$=\frac{(a;q)_\infty(-q;q)_\infty(aq^2/b^2;q^2)_\infty}{(aq/b;q)_\infty(-q/b;q)_\infty(a;q^2)_\infty}\text{ (by Thm. 1.64)}$$

$$=\frac{(a;q)_\infty(-q;q)_\infty}{(aq/b;q)_\infty(-q/b;q)_\infty}.$$

□

Theorem 1.49 (*q*-Pfaff–Saalschütz).

$$_3\phi_2\left[\begin{matrix}a,b,q^{-n}\\c,abc^{-1}q^{1-n}\end{matrix};q,q\right]=\frac{(c/a;q)_n(c/b;q)_n}{(c;q)_n(c/ab;q)_n}.$$

Proof. The *q*-WZ proof certificate is

$$R_q(n,k)=\frac{\left(1-aq^k\right)\left(1-bq^k\right)\left(1-q^{n-k}\right)}{(1-q^n)(1-cq^{n-1})\left(1-\frac{abq^{k-n+1}}{c}\right)}.$$

□

Theorem 1.50 (Jackson's $_6\phi_5$ sum).

$$_6\phi_5\left[\begin{matrix}a,q\sqrt{a},-q\sqrt{a},b,c,q^{-n}\\\sqrt{a},-\sqrt{a},aq/b,aq/c,aq^{n+1}\end{matrix};q,\frac{aq^{n+1}}{bc}\right]=\frac{(aq;q)_n(aq/bc;q)_n}{(aq/b;q)_n(aq/c;q)_n}.$$

Proof. The *q*-WZ proof certificate is

$$R_q(n,k)=\frac{aq^n\left(1-aq^k\right)\left(1-bq^k\right)\left(1-cq^k\right)\left(1-q^{n-k}\right)}{bc(1-q^n)(1-aq^{2k})\left(1-\frac{aq^n}{b}\right)\left(1-\frac{aq^n}{c}\right)}.$$

□

1.7   *q*-Binomial Theorem and Jacobi's Triple Product Identity

Definition 1.51. The *q-binomial coëfficient* is given by

$$\binom{n}{k}_q:=\begin{cases}\dfrac{(q;q)_n}{(q;q)_k(q;q)_{n-k}}=\dfrac{[n]_q!}{[k]_q![n-k]_q!}&\text{if }0\le k\le n,\\0&\text{otherwise.}\end{cases}$$

Although the definition makes it appear to be a rational function, the q-binomial coëfficient $\binom{n}{k}_q$ is actually a polynomial in q of degree $k(n-k)$, and is often referred to as the *Gaussian polynomial*, just as the ordinary binomial coëfficient is defined as a ratio of factorials, but actually turns out to be an integer. It is immediate from the definition that

$$\binom{n}{k}_q = \binom{n}{n-k}_q. \tag{1.49}$$

Theorem 1.52. *The q-binomial coëfficients satisfy the following q-Pascal triangle recurrences:*

$$\binom{n}{k}_q = \binom{n-1}{k}_q + q^{n-k}\binom{n-1}{k-1}_q, \tag{1.50}$$

$$\binom{n}{k}_q = q^k\binom{n-1}{k}_q + \binom{n-1}{k-1}_q. \tag{1.51}$$

Proof. By definition, for $n \geq k > 0$, the right hand side of (1.50) is

$$\binom{n-1}{k}_q + q^{n-k}\binom{n-1}{k-1}_q = \frac{(q;q)_{n-1}}{(q;q)_k(q;q)_{n-k-1}} + q^{n-k}\frac{(q;q)_{n-1}}{(q;q)_{k-1}(q;q)_{n-k}}$$

$$= \frac{(1-q^{n-k})(q;q)_{n-1} + q^{n-k}(1-q^k)(q;q)_{n-1}}{(q;q)_k(q;q)_{n-k}}$$

$$= \frac{(1-q^{n-k}+q^{n-k}-q^n)(q;q)_{n-1}}{(q;q)_k(q;q)_{n-k}}$$

$$= \frac{(q;q)_n}{(q;q)_k(q;q)_{n-k}}$$

$$= \binom{n}{k}_q.$$

Eq. (1.50) reduces to $1 = 1 + 0$ in the case $n \geq k = 0$, and to $0 = 0 + 0$ in the case $k > n$. Eq. (1.51) can be proved similarly. ☐

In order to fully demonstrate the symmetry of the q-binomial coëfficient, we sometimes write

$$\binom{n}{k, n-k}_q$$

instead of the more brief $\binom{n}{k}_q$.

We remark that q-binomial coëfficient

$$\binom{k+l}{k, l}_q$$

encodes in the exponents of q the partitions of length at most l into parts at most k.

For example, while we may use the definition of the q-binomial coëfficient to calculate that

$$\binom{6}{3,3}_q = \frac{(1-q^6)(1-q^5)(1-q^4)}{(1-q^3)(1-q)(1-q^2)} = (1+q^3)(1+q+q^2+q^3+q^4)(1+q^2)$$

$$= 1 + q + 2q^2 + 3q^3 + 3q^5 + 3q^6 + 2q^7 + q^8 + q^9,$$

it is more enlightening to observe that

$$\binom{3+3}{3,3}_q = 1 + q^1 + (q^2 + q^{1+1}) + (q^3 + q^{2+1} + q^{1+1+1}) + (q^{3+1} + q^{2+2} + q^{2+1+1})$$

$$+ (q^{3+2} + q^{3+1+1} + q^{2+2+1}) + (q^{3+3} + q^{3+2+1} + q^{2+2+2})$$

$$+ (q^{3+3+1} + q^{3+2+2}) + q^{3+3+2} + q^{3+3+3}.$$

Let us revisit the q-Pascal triangle recurrence (1.51) with $n = k + l$:

$$\binom{k+l}{k,l}_q = q^k \binom{k+l-1}{k,l-1}_q + \binom{k+l-1}{k-1,l}_q. \tag{1.52}$$

The left-hand side generates partitions of length at most l with all parts at most k. The right-hand side splits these partitions into two classes: $q^k \binom{k+l-1}{k,l-1}_q$ generates those partitions with largest part exactly k, and the other term $\binom{k+l-1}{k-1,l}_q$ generates those partitions with largest part strictly less than k.

Theorem 1.53. *The q-binomial coëfficient*

$$\binom{k+l}{k}_q = \sum_{n=0}^{kl} p_{k,l}(n) q^n,$$

where $p_{k,l}(n)$ denotes the number of partitions of n with length $\leq l$ and each part at most k.

Proof. Left as an exercise in light of the discussion above. □

Remark 1.54. The q-binomial coëfficient $\binom{n}{k}_q$ equals the number of k-dimensional subspaces of an n-dimensional vector space over the finite field $GF(q)$, where $q = p^n$ for some prime p and positive integer n. See [And76, Chapter 13] for more details.

The following simple proposition will prove useful later.

Proposition 1.55.

$$\binom{n}{k}_{1/q} = q^{k(k-n)} \binom{n}{k}_q. \tag{1.53}$$

Proof.

$$\binom{n}{k}_{1/q} = \frac{(1/q; 1/q)_n}{(1/q; 1/q)_k (1/q; 1/q)_{n-k}}$$

$$= \frac{(-1)^n q^{-\binom{n+1}{2}}(q; q)_n}{(-1)^k q^{-\binom{k+1}{2}}(q; q)_k (-1)^{n-k} q^{-\binom{n-k+1}{2}}(q; q)_{n-k}}$$

$$= q^{k(k-n)} \frac{(q; q)_n}{(q; q)_k (q; q)_{n-k}}$$

$$= q^{k(k-n)} \binom{n}{k}_q.$$

$\square$

We note the asymptotic results

$$\lim_{n \to \infty} \binom{n}{k}_q = \frac{1}{(q; q)_k} \tag{1.54}$$

and

$$\lim_{n \to \infty} \binom{2n+a}{n+b}_q = \frac{1}{(q; q)_\infty} \tag{1.55}$$

for any a and b.

1.7.1 Finite q-Binomial Theorem

Theorem 1.56 (Finite q-binomial theorem).

$$\sum_{k=0}^{n} \binom{n}{k}_q (-1)^k z^k q^{k(k-1)/2} = (z; q)_n. \tag{1.56}$$

Proof. After rewriting (1.56) in the form

$$\sum_{k=0}^{n} \binom{n}{k}_q \frac{(-1)^k z^k q^{k(k-1)/2}}{(z; q)_n} = 1,$$

the q-WZ certificate is

$$R_q(n, k) = \frac{z q^{n-1}(1 - q^{n-k})}{1 - q^n}.$$

$\square$

Historical Note 1.57. Theorem 1.56 is the oldest known incarnation of the q-binomial theorem. It appeared in a book by Heinrich August Rothe in 1811 [Rot11]. Ironically, it only appeared in the preface where the author was complaining that his publisher had forced him to remove his most interesting results and this theorem was given as an example of such a result. One is left to wonder what other treasures Rothe may have had at the time.

Remark 1.58. To recover the ordinary binomial theorem,

$$\sum_{k=0}^{n} \binom{n}{k} x^k y^{n-k} = (x+y)^n,$$

set $z = -x/y$ in (1.56), multiply both sides by y^n, and send $q \to 1$.

1.7.2 Jacobi's Triple Product Identity and Ramanujan's Theta Functions

The following very important result, known as Jacobi's triple product identity, was once thought to be much deeper than the q-binomial theorem. Next, we show that it is in fact a fairly simple corollary of it.

Corollary 1.59 (Jacobi's triple product identity).

$$\sum_{j=-\infty}^{\infty} (-1)^j q^{j(j-1)/2} x^j = (x;q)_\infty (q/x;q)_\infty (q;q)_\infty. \qquad (1.57)$$

Proof. Let $n = 2m$ in Theorem 1.56 and set $j = k - m$ to obtain

$$(z;q)_{2m} = \sum_{j=-m}^{m} \binom{2m}{m+j}_q (-1)^{j+m} q^{(j+m)(j+m-1)/2} z^{j+m}.$$

Then set $z = xq^{-m}$ and rewrite $(xq^{-m};q)_{2m}$ as

$$(xq^{-m};q)_m (x;q)_m = (-1)^m x^m q^{-m^2+m(m-1)/2} (q/x;q)_m (x;q)_m.$$

The above identity then becomes

$$(q/x;q)_m (x;q)_m = \sum_{j=-m}^{m} \frac{(q;q)_{2m}(-1)^j q^{j(j-1)/2} x^j}{(q;q)_{m+j}(q;q)_{m-j}}.$$

Finally, send $m \to \infty$, and the result follows. □

By setting $x = zq^{1/2}$ and $q = w^2$ in (1.57), we obtain another useful form of Jacobi's triple product identity:

$$\sum_{j=-\infty}^{\infty} (-1)^j w^{j^2} z^j = (wz^{-1};w^2)_\infty (wz;w^2)_\infty (w^2;w^2)_\infty. \qquad (1.58)$$

Remark 1.60. By setting $w = q^{3/2}$ and $z = q^{1/2}$ in (1.58), Euler's pentagonal numbers theorem, Theorem 1.19, is recovered.

At this point, let us introduce some notation due to Ramanujan [AB05, p. 9, Eq. (1.1.5)]. *Ramanujan's theta function $f(a,b)$ is given by*

$$f(a,b) := \sum_{n=-\infty}^{\infty} a^{n(n+1)/2} b^{n(n-1)/2}. \tag{1.59}$$

It is called a "theta" function, even though "ϑ" does not appear in the notation, since one can change variables to show its equivalence to the elliptic theta functions of Jacobi:

$$\vartheta_1(v|\tau) := -i \sum_{n=-\infty}^{\infty} (-1)^n \exp\left\{\pi i\left[\tau\left(n+\frac{1}{2}\right)^2 + (2n+1)v\right]\right\} \tag{1.60}$$

$$\vartheta_2(v|\tau) := \sum_{n=-\infty}^{\infty} \exp\left\{\pi i\left[\tau\left(n+\frac{1}{2}\right)^2 + (2n+1)v\right]\right\} \tag{1.61}$$

$$\vartheta_3(v|\tau) := \sum_{n=-\infty}^{\infty} \exp\left(\pi i(2vn+\tau n^2)\right), \tag{1.62}$$

$$\vartheta_4(v|\tau) := \sum_{n=-\infty}^{\infty} (-1)^n \exp\left(\pi i(2vn+\tau n^2)\right), \tag{1.63}$$

where we have presented the definitions equivalent to those in Tannery and Molk [TM96, vol. 2, p. 252], and Rademacher [Rad73, p. 166]. In [TM96] and [Rad73], however, the definitions are given with the abbreviations $q := e^{\pi i \tau}$ and $z := e^{2\pi i v}$, so one must be careful as it is more common in the current literature to let $q := e^{2\pi i \tau}$, as we did earlier when discussing Dedekind's eta function. It is also common (though by no means universal) in the contemporary literature to use z in place of τ, which is clearly incompatible with Rademacher's use of $z = e^{2\pi i v}$. This is why we have chosen to present the theta functions above directly as functions of v and τ, perhaps at the expense of the ease of readability offered by introducing q and z. Superficially different definitions of the theta functions appear elsewhere in the literature, e.g. Whittaker and Watson [WW27, p. 463]. It should be noted that the theta functions satisfy a functional equation analogous to (1.10):

$$\vartheta_1\left(\frac{v}{c\tau+d}\bigg|\frac{a\tau+b}{c\tau+d}\right) = -i\epsilon(a,b,c,d)^3\sqrt{\frac{c\tau+d}{i}}e^{\pi i c v^2/(c\tau+d)}\vartheta_1(v\mid\tau),$$

$$\tag{1.64}$$

for $c > 0$ and

$$\vartheta_1(v\mid\tau+b) = e^{\pi i b/4}\vartheta_1(v\mid\tau), \tag{1.65}$$

in the $c = 0$ case, where $\epsilon(a,b,c,d)$ is defined in (1.11).

Observe the equivalence of $\vartheta_1(v|\tau)$ with $f(a,b)$:

$$\vartheta_1(v\mid\tau) = -ie^{i\pi\tau/4}e^{\pi i v}\sum_{n=-\infty}^{\infty}(-1)^n e^{\pi i\tau(n^2+n)}e^{2\pi i v n}$$

$$= -ie^{\pi i\tau/4}e^{\pi iv} \sum_{n=-\infty}^{\infty} (-e^{-2\pi iv})^{n(n-1)/2}(-e^{2\pi i(v+\tau)})^{n(n+1)/2}$$

$$= e^{\pi i(v+\tau/4-1/2)} f(-e^{-2\pi iv}, -e^{2\pi i(v+\tau)}).$$

In particular, we shall primarily be interested in theta functions of the following form:

$$\vartheta_1(\alpha\tau \mid (\alpha+\beta)\tau) = ie^{\pi i\tau(\beta-3\alpha)/4} f(-x^\alpha, -x^\beta),$$

where here we are following Rademacher [Rad73] and letting $x := \exp(2\pi i\tau)$.

Ramanujan's notation is particularly convenient for the subject matter of this work, as we shall explain below.

Jacobi's triple product identity, in Ramanujan's notation, is given by

$$f(a, b) = (-a; ab)_\infty(-b; ab)_\infty(ab; ab)_\infty. \tag{1.66}$$

Certain special cases of $f(a, b)$ arose so often in Ramanujan's work that he defined the following special notations for them:

$$\varphi(q) := f(q, q), \tag{1.67}$$

$$\psi(q) := f(q, q^3), \tag{1.68}$$

$$f(-q) := f(-q, -q^2). \tag{1.69}$$

We record the following immediate corollary of (1.66).

Corollary 1.61.

$$f(-q) = (q; q)_\infty, \tag{1.70}$$

$$\varphi(-q) = \frac{(q; q)_\infty}{(-q; q)_\infty}, \tag{1.71}$$

$$\psi(-q) = \frac{(q^2; q^2)_\infty}{(-q; q^2)_\infty}. \tag{1.72}$$

The proofs are not difficult but require facility with notation that may be unfamiliar to the beginner. As the reader will need to become comfortable with the q-Pochhammer and Ramanujan theta notations in order to fully appreciate what is to come, we leave the proofs as exercises.

Subsequently, we shall freely use Ramanujan's notations $f(a, b)$, and the abbreviations $\varphi(q)$ and $\psi(q)$, but refrain from using Ramanujan's abbreviation $f(-q)$, preferring instead $(q; q)_\infty$, so as to avoid any possible confusion between $f(a, b)$ and $f(-q)$.

Notice that Eq. (1.70) is a restatement of Euler's pentagonal numbers theorem. Accordingly, Eqs. (1.71) and (1.72) are due to Gauss, and may be appropriately named "Gauss' square numbers theorem," and "Gauss' hexagonal numbers theorem" respectively.

A word on the convenience of Ramanujan's "f" notation. For our purposes, the arguments of f will always be plus or minus a power of q. Since

$$f(-q^\alpha, -q^\beta) = (q^\alpha; q^{\alpha+\beta})_\infty (q^\beta; q^{\alpha+\beta})_\infty (q^{\alpha+\beta}; q^{\alpha+\beta})_\infty$$
$$= 1 - q^\alpha - q^\beta + q^{3\alpha+\beta} + q^{\alpha+3\beta} - \cdots,$$

both the infinite product representation (in terms of q-Pochhammer symbols) and the first few terms of the series expansion can be read off directly from the notation.

1.7.3 Quintuple Product Identity

The quintuple product identity has been discovered and independently rediscovered many times over the past century. The earliest known appearance in the literature appears to be due to Robert Fricke in 1916 [Fri16, p. 432, Eq. (6)], although Slater [Sla54] and D. B. Sears [Sea52] were able to show that the quintuple product identity can actually be deduced from the three term relation for sigma functions known to Weierstrass in the nineteenth cenutry. Ramanujan independently discovered the quintuple product identity no later than 1919, and recorded it in his lost notebook in the following form [AB09, p. 54, Entry 3.1.1].

Theorem 1.62 (Quintuple product identity).

$$\frac{f(-\lambda x, -x^2)}{f(-x, -\lambda x^2)} f(-\lambda x^3, -\lambda^2 x^6) = f(-\lambda^2 x^3, -\lambda x^6) + x f(-\lambda, -\lambda^2 x^9). \quad (1.73)$$

A more symmetrical presentation of the quintuple product identity is

$$f(-wx^3, -w^2 x^{-3}) + x f(-wx^{-3}, -w^2 x^3)$$
$$= (-w/x, -x, w; w)_\infty (w/x^2, wx^2; w^2)_\infty, \quad (1.74)$$

where the product of five infinite rising q-factorials clearly appears on the right side.

For a history of the quintuple product identity and many proofs and generalizations, see Shaun Cooper's article [Coo06]. We will derive Theorem 1.62 as a corollary of the following finite form due to Peter Paule [Pau94].

Theorem 1.63 (Paule's finite analog of the quintuple product identity).

$$\sum_{k=-n}^{n} (-1)^k q^{k(3k+1)/2} x^{3k} \binom{2n}{n+k}_q \frac{1 - x^2 q^{2k+1}}{(x^2 q^2; q)_{n+k} (x^{-2}; q)_{n-k}} = \frac{1 - x^2 q}{(xq; q)_n (x^{-1}; q)_n}. \quad (1.75)$$

Proof. Each side of (1.75) is annihilated by the operator

$$A(\eta) := \eta^0 - (1 - q^{n-1}/x)^{-1} (1 - xq^n)^{-1} \eta^{-1};$$

this is certified on the left side by

$$R_q(n,k) = \frac{xq^{n+2k}(1 - x^{-2}q^{n-k-1})(1 - xq^{k+1})}{(1 - q^{n-1}/x)(1 - x^2q^{2k+1})(1 - xq^n)}$$
$$\times \frac{[n-k]_q([n-k-1]_q - xq^n[2n-1]_q + x^2q[n+k]_q)}{[2n]_q[2n-1]_q}.$$

Both sides evaluate to $1 - x^2q$ at $n = 0$. □

Note that upon letting $n \to \infty$ in (1.75), one obtains

$$\sum_{k\in\mathbb{Z}}(-1)^k q^{k(3k+1)/2} x^{3k}(1 - x^2q^{2k+1}) = \frac{(q;q)_\infty(x^{-2};q)_\infty(x^2q;q)_\infty}{(x^{-1};q)_\infty(xq;q)_\infty}, \quad (1.76)$$

which is the form of the quintuple product identity found by G. N. Watson.

Exercise. Perform the appropriate changes of variables to show the equivalence of the various forms of the quintuple product identity presented here.

1.7.4 q-Binomial Series

Theorem 1.64 (q-binomial series).

$$\sum_{n=0}^{\infty} \frac{(a;q)_n}{(q;q)_n} z^n = \frac{(az;q)_\infty}{(z;q)_\infty}. \quad (1.77)$$

Proof. Let

$$F(z) := \frac{(az;q)_\infty}{(z;q)_\infty} = \sum_{n=0}^{\infty} A_n z^n.$$

From the representation of $F(z)$ as a ratio of infinite products, it is easy to see that

$$(1-z)F(z) = (1-az)F(zq).$$

But this means

$$(1-z)\sum_{n=0}^{\infty} A_n z^n = (1-az)\sum_{n=0}^{\infty} A_n q^n z^n,$$

or

$$\sum_{n=0}^{\infty} A_n z^n - \sum_{n=1}^{\infty} A_{n-1} z^n = \sum_{n=0}^{\infty} A_n q^n z^n - \sum_{n=1}^{\infty} A_{n-1} aq^{n-1} z^n. \quad (1.78)$$

Comparing coëfficients of z^n on both sides of (1.78), we obtain, for $n \geq 1$,

$$A_n - A_{n-1} = q^n A_n - aq^{n-1} A_{n-1},$$

which is equivalent to

$$A_n = \frac{1 - aq^{n-1}}{1 - q^n} A_{n-1}. \tag{1.79}$$

Starting with $A_0 = 1$, and iterating (1.79), we obtain

$$A_n = \frac{(a;q)_n}{(q;q)_n},$$

and the result follows. □

For later use, we shall record the following simple but very useful corollary, which is obtained by setting $a = q^j$ in Eq. (1.77).

Corollary 1.65.

$$\sum_{n=0}^{\infty} \binom{n+j-1}{n}_q t^n = \frac{1}{(t;q)_j}. \tag{1.80}$$

1.7.5 Exercises

1. Prove that $\binom{n}{k}_q$ is a polynomial in q of degree $k(n-k)$.

2. Prove Eq. (1.51).

3. Prove Eqs. (1.70), (1.71), and (1.72) from (1.66).

4. Given that (1.70) is "Euler's pentagonal numbers theorem," why would it be appropriate to name Eqs. (1.71) and (1.72) "Gauss' square numbers theorem" and "Gauss' hexagonal numbers theorem" respectively?

5. Obtain Eq. (1.5) as a corollary of Theorem 1.64.

6. Find the appropriate change of variables to rewrite Ramanujan's $f(a,b)$ in terms of Jacobi's $\vartheta_\lambda(v|\tau)$, and vice versa, for $\lambda = 1, 2, 3, 4$.

1.8 q-Trinomial Coëfficients

1.8.1 Ordinary Trinomial Coëfficients

In our studies thus far, we have encountered the binomial theorem (Theorem 1.36) and its q-analog (Theorem 1.56). It happens that trinomial coëfficients and several of their q-analogs are also of importance in the theory of partitions.

Just as binomial coëfficients arise in the expansion of $(x+1)^n$:

$$(1+x)^n = \sum_{j=0}^{n} \binom{n}{j} x^j,$$

the trinomial coëfficients arise in the expansion of $(x+1+x^{-1})^n$:

$$(1+x+x^{-1})^n = \sum_{j=-n}^{n} \left(\!\!\binom{n}{j}\!\!\right) x^j. \tag{1.81}$$

The trinomial coëfficients can be arranged into a triangle analogous to Pascal's triangle:

$$\begin{array}{ccccccccc}
 & & & & 1 & & & & \\
 & & & 1 & 1 & 1 & & & \\
 & & 1 & 2 & 3 & 2 & 1 & & \\
 & 1 & 3 & 6 & 7 & 6 & 3 & 1 & \\
1 & 4 & 10 & 16 & 19 & 16 & 10 & 4 & 1
\end{array}$$
$$\vdots$$

Notice the symmetry

$$\left(\!\!\binom{n}{j}\!\!\right) = \left(\!\!\binom{n}{-j}\!\!\right) \tag{1.82}$$

and the apparent recurrence rule

$$\left(\!\!\binom{n}{j}\!\!\right) = \left(\!\!\binom{n-1}{j-1}\!\!\right) + \left(\!\!\binom{n-1}{j}\!\!\right) + \left(\!\!\binom{n-1}{j+1}\!\!\right). \tag{1.83}$$

Observe that

$$\sum_{j=-n}^{n} \left(\!\!\binom{n}{j}\!\!\right) x^j = (x+1+x^{-1})^n = x^{-n}\Big(1+x(1+x)\Big)^n$$

$$= x^{-n} \sum_{k=0}^{n} \binom{n}{k} (x+x^2)^{n-k}$$

$$= \sum_{k=0}^{n} \binom{n}{k} \sum_{r=0}^{n-k} \binom{n-k}{r} x^{2r} x^{-k-r}$$

$$= \sum_{k=0}^{n} \sum_{r=0}^{n-k} \binom{n}{k} \binom{n-k}{r} x^{-k+r}$$

$$= \sum_{k=0}^{n} \sum_{j=-k}^{n-k} \binom{n}{k} \binom{n-k}{k+j} x^j$$

$$= \sum_{j=-n}^{n} \sum_{k=0}^{n} \binom{n}{k} \binom{n-k}{k+j} x^j,$$

and so

$$\left(\!\!\binom{n}{j}\!\!\right) = \sum_{k=0}^{n} \binom{n}{k}\binom{n-k}{k+j}. \tag{1.84}$$

Next observe that

$$\sum_{j=-n}^{n} \left(\!\!\binom{n}{j}\!\!\right) x^{j+n} = \left((1+x)^2 - x\right)^n$$

$$= \sum_{j=-n}^{n}\sum_{k\geq 0}(-1)^k \binom{n}{k}\binom{2n-2k}{n-j-k} x^{j+n},$$

and so we also have

$$\left(\!\!\binom{n}{j}\!\!\right) = \sum_{k\geq 0}(-1)^k \binom{n}{k}\binom{2n-2k}{n-j-k}. \tag{1.85}$$

1.8.2 Several q-Trinomial Coëfficients

In [AB87], Andrews and Rodney J. Baxter introduced the following q-analog of (1.84):

$$\left(\!\!\binom{n}{j}\!\!\right)_q := \sum_{k=0}^{n} q^{k(k+j)} \binom{n}{k}_q \binom{n-k}{k+j}_q. \tag{1.86}$$

We need to verify that the following recurrence holds.

Theorem 1.66 (q-analog of Eq. (1.83)). *For $n \geq 1$,*

$$\left(\!\!\binom{n}{j}\!\!\right)_q = q^{n+j}\left(\!\!\binom{n-1}{j+1}\!\!\right)_q + \left(\!\!\binom{n-1}{j}\!\!\right)_q + q^{n-j}\left(\!\!\binom{n-1}{j-1}\!\!\right)_q$$

$$+ q^{n-1}(1-q^{n-1})\left(\!\!\binom{n-2}{j}\!\!\right)_q. \tag{1.87}$$

Proof. Let

$$T(n,j,k) := q^{k(k+j)} \binom{n}{k}_q \binom{n-k}{k+j}_q,$$

i.e. $T(n,j,k)$ is the expression that is summed over k in the right-hand side of Eq. (1.86). Then

$$T(n-1,j,k) + q^{n+j}T(n-1,j+1,k-1) + q^{n-j}T(n-1,j-1,k)$$
$$+ q^{n-1}(1-q^{n-1})T(n-2,j,k-1)$$
$$= \frac{q^{k(k+j)}}{1-q^n}\binom{n}{k}_q\binom{n-k}{j+k}_q \left\{(1-q^{n-2k-j}) + q^{n-k}(1-q^k) + q^{n-j-k}(1-q^{j+k})\right\}$$

$$+q^{n-2k-j}(1-q^k)(1-q^{j+k})\}$$

$$= q^{k(k+j)}\binom{n}{k}_q\binom{n-k}{j+k}_q$$

$$= T(n,j,k). \tag{1.88}$$

The result follows by summing the extremes of (1.88) over k. $\qquad\square$

Theorem 1.67.

$$\lim_{n\to\infty}\left(\!\binom{n}{j}\!\right)_q = \frac{1}{(q;q)_\infty}.$$

Proof.

$$\lim_{n\to\infty}\left(\!\binom{n}{j}\!\right)_q = \sum_{k=0}^\infty \frac{q^{k(k+j)}}{(q;q)_k(q;q)_{k+j}}$$

$$= \frac{1}{(q;q)_\infty} \quad \text{(by (1.96))}.$$

$\qquad\square$

Define

$$\mathscr{T}_n(x,q) := \sum_{j=-n}^{n} x^j q^{j(3j-1)/2}\left(\!\binom{n}{j}\!\right)_q^3. \tag{1.89}$$

Notice that $\mathscr{T}_n(x,1)$ is the right hand side of Eq. (1.81).

Next, apply Eq. (1.87) to each term of (1.89) to find

$$\mathscr{T}_n(x,q) = (1+xq^{3n-2}+x^{-1}q^{3n-1})\mathscr{T}_{n-1}(x,q) + q^{3n-3}(1-q^{3n-3})\mathscr{T}_{n-2}(x,q) \tag{1.90}$$

and observe $\mathscr{T}_0(x,q)=1$, $\mathscr{T}_1(x,q)=1+xq+x^{-1}q^2$.

So, we find $\mathscr{T}_n(x,1) = (1+x+x^{-1})\mathscr{T}_{n-1}(x,1)$, for $n\geq 1$, and thus $\mathscr{T}_n(x,1) = (1+x+x^{-1})^n$. Upon equating the two presentations of $\mathscr{T}_n(x,q)$, we thus have a q-analog of the trinomial theorem (1.81).

As it turns out, $\left(\!\binom{n}{j}\!\right)_q$ is not the only known useful q-analog of $\binom{n}{j}$. Two further q-analogs, based on the presentation (1.85), are

$$T_0(n,j,q) := \sum_{k=0}^{n}(-1)^k\binom{n}{k}_{q^2}\binom{2n-2k}{n-j-k}_q, \tag{1.91}$$

$$T_1(n,j,q) := \sum_{k=0}^{n}(-q)^k\binom{n}{k}_{q^2}\binom{2n-2k}{n-j-k}_q. \tag{1.92}$$

1.8.3   Exercises

1. Prove Eq. (1.82).

2. Prove Eq. (1.83).

3. Prove Eq. (1.85).

1.9 q-Hypergeometric Transformation Formulæ

We have already examined some classical examples of $(q\text{-})$hypergeometric series that sum to a finite or infinite product. Transformation formulæ are also central to the theory of $(q\text{-})$hypergeometric series.

1.9.1   Heine–Rogers Transformations

Heine proved the following $_2\phi_1$ transformation formula

$$_2\phi_1\left[\begin{matrix} a,b \\ c \end{matrix}; q, z\right] = \frac{(b;q)_\infty (az;q)_\infty}{(c;q)_\infty (z;q)_\infty} \, _2\phi_1\left[\begin{matrix} c/b, z \\ az \end{matrix}; q, b\right] \qquad (1.93)$$

which is the limiting $n \to \infty$ case of the next theorem, due to Andrews [And09, p. 3, Theorem 2]

Theorem 1.68 (Andrews' finite Heine transformation).

$$_3\phi_2\left[\begin{matrix} a,b,q^{-n} \\ c, q^{1-n}/t \end{matrix}; q, q\right] = \frac{(b;q)_n (at;q)_n}{(c;q)_n (t;q)_n} \, _3\phi_2\left[\begin{matrix} c/b, z, q^{-n} \\ at, q^{1-n}/b \end{matrix}; q, q\right] \qquad (1.94)$$

Proof. As an identity of finite q-hypergeometric forms, we may prove this automatically using the q-Zeilberger algorithm.

Denote the left-hand side of (1.94) by $f(n)$. The q-Zeilberger algorithm tells us that $f(n)$ satisfies the recurrence

$$
\begin{aligned}
&f(n) \\
&+ \frac{\left(-abtq^{2n} + atq^{n+1} + btq^{n+1} - ctq^{2n} + cq^{n+1} + q^{n+2} - q^3 - q^2\right)}{q^2\left(1 - cq^{n-1}\right)\left(1 - tq^{n-1}\right)} f(n-1) \\
&+ \frac{q\left(1 - q^{n-1}\right)\left(1 - atq^{n-2}\right)\left(1 - btq^{n-2}\right)}{\left(1 - cq^{n-1}\right)\left(1 - tq^{n-2}\right)\left(1 - tq^{n-1}\right)} f(n-2) = 0, \quad (1.95)
\end{aligned}
$$

certified by the rational function

$$-\frac{t(q-t)\left(aq^k - 1\right)\left(bq^k - 1\right)q^{-k+2n+1}\left(q^n - q^k\right)}{\left(q^n - 1\right)\left(q - cq^n\right)\left(q^{k+1} - tq^n\right)\left(q^{k+2} - tq^n\right)}.$$

The q-Zeilberger algorithm shows that the right-hand side of (1.94) also satisfies the recurrence (1.95), as certified by the rational function

$$\frac{(b-q)q^{-k+2n+1}\left(q^n - q^k\right)\left(tq^k - 1\right)\left(cq^k - b\right)}{\left(q^n - 1\right)\left(q - atq^n\right)\left(q^{k+1} - bq^n\right)\left(q^{k+2} - bq^n\right)}.$$

Upon verifying that both sides agree at $n = 0$ and 1, the result follows. $\square$

Rogers noticed that the left side of (1.93) is clearly symmetric in a and b, yet on the right side, the symmetry is not obvious. Taking advantage of this apparent asymmetry, he was able to derive [Rog93b] the following variants of (1.93):

$$
{}_2\phi_1\left[\begin{matrix} a, b \\ c \end{matrix}; q, t\right] = \frac{(c/b; q)_\infty (bt; q)_\infty}{(t; q)_\infty (c; q)_\infty} {}_2\phi_1\left[\begin{matrix} b, abt/c \\ bz \end{matrix}; q, c/b\right], \tag{1.96}
$$

$$
{}_2\phi_1\left[\begin{matrix} a, b \\ c \end{matrix}; q, t\right] = \frac{(abt/c; q)_\infty}{(t; q)_\infty} {}_2\phi_1\left[\begin{matrix} c/a, c/b \\ c \end{matrix}; q, abt/c\right]. \tag{1.97}
$$

1.9.2   Watson's q-Analog of Whipple's Theorem

One of the most important formulæ in the theory of q-hypergeometric series transforms a very-well-poised ${}_8\phi_7$ to a balanced ${}_4\phi_3$, and is due to G. N. Watson [Wat29].

Theorem 1.69 (Watson's q-analog of Whipple's theorem).

$$
{}_8\phi_7\left[\begin{matrix} a, q\sqrt{a}, -q\sqrt{a}, b, c, d, e, q^{-n} \\ \sqrt{a}, -\sqrt{a}, aq/b, aq/c/aq/d/aq/e, aq^{n+1} \end{matrix}; q, \frac{a^2 q^{n+2}}{bcde}\right]
$$

$$
= \frac{(aq; q)_n (aq/de; q)_n}{(aq/d; q)_n (aq/e; q)_n} {}_4\phi_3\left[\begin{matrix} q^{-n}, d, e, aq/bc \\ aq/b, aq/c, deq^{-n}/a \end{matrix}; q, q\right]. \tag{1.98}
$$

Proof. Let $f(n)$ denote the left-hand side of (1.98). The q-Zeilberger algorithm reveals that $f(n)$ satisfies the second-order recurrence

$$
f(n) = \frac{(aq^n - 1)(q^n - q)(aq^n - q)(bcde - a^2 q^n)}{q(b - aq^n)(c - aq^n)(d - aq^n)(e - aq^n)} f(n-2)
$$

$$
+ \frac{aq^n - 1}{q(b - aq^n)(c - aq^n)(d - aq^n)(e - aq^n)}
$$

$$
\times \left(a^3(q+1)q^{3n} + q^{n+1}\left(a^2 q + a(bec + bed + bcd + cde) + bcde\right)\right.
$$

$$
\left. -aq^{2n}(aq(a + b + c + d + e) + bcde) - bcde(q+1)q\right) f(n-1) \tag{1.99}
$$

as certified by the rational function

$$
\frac{a^2\left(aq^k - 1\right)\left(bq^k - 1\right)\left(cq^k - 1\right)\left(dq^k - 1\right)\left(eq^k - 1\right)q^{2n-2k}\left(q^n - q^k\right)}{(q^n - 1)(aq^{2k} - 1)(b - aq^n)(c - aq^n)(d - aq^n)(e - aq^n)}.
$$

Again by the q-Zeilberger algorithm, the right hand side of (1.98) also satisfies the recurrence (1.99), as certified by

$$
-\frac{a(a - de)\left(dq^k - 1\right)\left(eq^k - 1\right)q^{2n-k}\left(q^n - q^k\right)\left(aq^{k+1} - bc\right)}{(q^n - 1)(b - aq^n)(c - aq^n)(deq^k - aq^n)(deq^{k+1} - aq^n)}.
$$

Upon checking that both sides agree at $n = 0$, and $n = 1$, the result follows.

$\square$

Remark 1.70. Theorem 1.69 is a q-analog of

$$_7F_6\left[\begin{matrix}a,1+\frac{a}{2},b,c,d,e,-n\\\frac{a}{2},1+a-b,1+a-e,1+a+n\end{matrix};1\right]$$
$$=\frac{(1+a)_n(1+a-d-e)_n}{(1+a-d)_n(1+a-e)_n}{}_4F_3\left[\begin{matrix}1+a-b-c,d,e,-n\\1+a-b,1+a-c,d+e-a-n\end{matrix};1\right],$$

a result due to F. J. W. Whipple [Whi26]; cf. [Bai35, p. 25].

1.10 Classical q-Series Identities in Two and Three Variables

The results given below are corollaries of some of the more general results given earlier, but as they possess significant combinatorial and analytic interest in their own right, they are labeled as theorems.

Theorem 1.71 (Euler).

$$\sum_{n=0}^{\infty}\frac{z^n q^{n(n-1)/2}}{(q;q)_n}=(-z;q)_\infty. \tag{1.100}$$

Proof. This is a corollary of Theorem 1.64, the q-binomial series. In Equation (1.77), replace z by $-z/a$ and let $a\to\infty$. □

Remark 1.72. Notice that Equation (1.100) gives two representations of $E_q(z/(1-q))$.

Theorem 1.73 (Cauchy).

$$\sum_{n=0}^{\infty}\frac{z^n q^{n^2+n}}{(q;q)_n(zq;q)_n}=\frac{1}{(zq;q)_\infty}. \tag{1.101}$$

Proof. In Theorem 1.47, set $c=z$ and let $a,b\to\infty$. □

The next identity is due to the number theorist Victor-Amédée Lebesgue (1791–1875) [Leb40]; not to be confused with the better known Henri Léon Lebesgue (1875–1941), famous for the Lebesgue integral.

Theorem 1.74 (V.-A. Lebesgue).

$$\sum_{n=0}^{\infty}\frac{q^{n(n+1)/2}(a;q)_n}{(q;q)_n}=\frac{(aq;q^2)_\infty}{(q;q^2)_\infty}. \tag{1.102}$$

Proof. Let $b\to\infty$ in Theorem 1.48. □

Theorem 1.75 (Limiting case of Heine's q-analog of Gauss' sum).

$$\sum_{n=0}^{\infty} \frac{(-1)^n c^n q^{n(n-1)/2}(a;q)_n}{a^n(q;q)_n(c;q)_n} = \frac{(c/a;q)_\infty}{(c;q)_\infty}. \tag{1.103}$$

Proof. Let $b \to \infty$ in Theorem 1.47. $\square$

The next two results are much more recent than those above; they first appeared in a 1973 paper by Andrews [And73].

Theorem 1.76 (Andrews' q-analog of Gauss' $_2F_1(\frac{1}{2})$ sum).

$$\sum_{n=0}^{\infty} \frac{(a;q)_n(b;q)_n q^{n(n+1)/2}}{(q;q)_n(abq;q^2)_n} = \frac{(aq;q^2)_\infty(bq;q^2)_\infty}{(q;q^2)_\infty(abq;q^2)_\infty}.$$

Theorem 1.77 (Andrews' q-analog of Bailey's $_2F_1(\frac{1}{2})$ sum).

$$\sum_{n=0}^{\infty} \frac{(b;q)_n(q/b;q)_n c^n q^{n(n-1)/2}}{(c;q)_n(q^2;q^2)_n} = \frac{(cq/b;q^2;q^2)_\infty(bc;q^2)_\infty}{(c;q)_\infty}.$$

Theorems 1.76 and 1.77 may be derived as follows. Start with D. B. Sears' identity [GR04, p. 360, Eq. (III.16)]

$$_4\phi_3\left[\begin{array}{c} q^{-n}, a, b, c \\ d, e, abcq^{1-n}/(de) \end{array} ; q, q\right]$$
$$= \frac{(b;q)_n(de/(ab);q)_n(de/(bc);q)_n}{(d;q)_n(e;q)_n(de/(abc);q)_n} {}_4\phi_3\left[\begin{array}{c} q^{-n}, d/b, e/b, de/(abc) \\ de/(ab), de/(bc), q^{1-n}/b \end{array} ; q, q\right], \tag{1.104}$$

which can be routinely verified using the q-Zeilberger algorithm. Then let n tend to infinity in (1.104) to obtain

$$_3\phi_2\left[\begin{array}{c} a, b, c \\ d, e \end{array} ; q, \frac{de}{abc}\right]$$
$$= \frac{(b;q)_\infty(de/(ab);q)_\infty(de/(bc);q)_\infty}{(d;q)_\infty(e;q)_\infty(de/(abc);q)_\infty} {}_3\phi_2\left[\begin{array}{c} d/b, e/b, de/(abc) \\ de/(ab), de/(bc) \end{array} ; q, b\right], \tag{1.105}$$

a formula due to Newman Hall [Hal36].

Next, specialize (1.105) appropriately to obtain an identity for which a limiting case simplifies to

$$\sum_{n=0}^{\infty} \frac{(\beta;q)_n(\alpha;q)_n(-1)^n q^{n(n-1)/2}(x\gamma/\beta)^n}{(q;q)_n(\gamma;q)_n(x\alpha;q)_n} = \frac{(x;q)_\infty}{(x\alpha;q)_\infty} \sum_{n=0}^{\infty} \frac{(\gamma/\beta;q)_n(\alpha;q)_n}{(q;q)_n(\gamma;q)_n} x^n. \tag{1.106}$$

Finally, both Theorems 1.77 and 1.76 follow by specializing (1.106) and then applying Theorem 1.48, Bailey's q-analog of Kummer's theorem.

Remark 1.78. A combinatorial interpretation and proof of Theorem 1.77, using overpartitions, is given by Sylvie Corteel and Jeremy Lovejoy in [CL09].

An *overpartition* of n is similar to a partition of n, except that the last occurrence of a given integer as a part might or might not be overlined. For example, in all there are eight overpartitions of 3:

$$(3) \quad (\bar{3}) \quad (21) \quad (\bar{2}1) \quad (2\bar{1}) \quad (\bar{2}\bar{1}) \quad (111) \quad (11\bar{1}).$$

The generating function for $\bar{p}(n)$, the number of overpartitions of n, is only slightly more complicated than that of $p(n)$,

$$\sum_{n\geq 0} \bar{p}(n)q^n = \frac{(-q;q)_\infty}{(q;q)_\infty},$$

and enjoys a rich theory that in many ways parallels the theory of (ordinary) partitions.

Overpartitions were formally introduced in a paper by Corteel and Lovejoy [CL04], and have subsequently spawned a significant literature.

Exercises. Fill in the details outlined in this section to derive Theorems 1.76 and 1.77.

1.11 Two-Variable Generating Functions

1.11.1 Generating Function for Partitions of n into l Parts

The preceding section saw the introduction of an additional formal variable beside q. This additional variable z can be given combinatorial significance. For instance, Euler's generating function for $p(n)$ may be easily generalized to

$$\sum_{n=0}^{\infty}\sum_{l=0}^{n} p(l,n)z^l q^n = \prod_{m\geq 1}\frac{1}{1-zq^m}, \qquad (1.107)$$

where $p(l,n)$ denotes the number of partitions of length l and weight n.
Alternatively, we could write

$$\sum_{\lambda \in \mathscr{P}} z^{\ell(\lambda)}q^{|\lambda|} = \frac{1}{(zq;q)_\infty},$$

where $\mathscr{P}$ denotes the set of all partitions.

The combinatorial significance of Equation (1.107) becomes transparent when the infinite product is written as

$$
\begin{aligned}
\frac{1}{1-zq} \cdot \frac{1}{1-zq^2} \cdot \frac{1}{1-zq^3} \cdots \\
= (1+z^1q^1 + z^2q^{1+1} + z^3q^{1+1+1} + \cdots) \\
\times (1+z^1q^2 + z^2q^{2+2} + z^3q^{2+2+2} + \cdots) \\
\times (1+z^1q^3 + z^2q^{3+3} + z^3q^{3+3+3} + \cdots)
\end{aligned}
$$

$$\vdots$$

1.11.2   Generating Function for Partitions of n into l Distinct Parts

By analogous reasoning, if we let $Q(l,n)$ denote the number of partitions into exactly l mutually distinct parts, we see that the corresponding two-variable generating function is $(1+zq)(1+zq^2)(1+zq^3)(1+zq^4)\cdots$, i.e. that

$$\sum_{n\geq 0}\sum_{l\geq 0} Q(l,n)z^l q^n = \prod_{m\geq 0}(1+zq^m), \qquad (1.108)$$

and thus if we wish to answer Naudé's question to Euler that opened the chapter ("How many ways can 50 be written as a sum of seven different positive integers?"), we need to extract the coëfficient of $z^7 q^{50}$ from the expansion of the infinite product in (1.108). Using a computer algebra system, the answer is quickly found to be 522.

To this point, the reader has seen that interpretation of q-series identities as generating functions can yield identities between restricted classes of partitions. It turns out that information can flow in the other direction as well; that is, insight into relationships between sets of partitions can lead to the discovery of q-series identities. The following is a case in point.

During his time on the faculty of Johns Hopkins University, Sylvester discovered and proved the identity [SF82, p. 281ff.]

$$\sum_{j\geq 0} \frac{z^j q^{j(3j-1)/2}(-zq;q)_{j-1}(1+zq^{2j})}{(q;q)_j} = (-zq;q)_\infty. \qquad (1.109)$$

Eq. (1.109) has the distinction of being the first q-series identity that was discovered and proved by combinatorial methods. (See the exercises following.)

In fact, Sylvester left it as an open challenge to find an algebraic/analytic proof of (1.109). Arthur Cayley rose to the challange [Cay84] and provided the following proof of (1.109).

Proof of Eq. (1.109). Let

$$F(z) := \sum_{j \geq 0} \frac{z^j q^{j(3j-1)/2}(-zq;q)_{j-1}}{(q;q)_j}(1+zq^{2j})$$

$$= \sum_{j \geq 0} \frac{z^j q^{j(3j-1)/2}(-zq;q)_{j-1}}{(q;q)_j}\left((1-q^j)+q^j(1+zq^j)\right)$$

$$= \sum_{j \geq 1} \frac{z^j q^{j(3j-1)/2}(-zq;q)_{j-1}}{(q;q)_{j-1}}$$

$$+ \sum_{j \geq 0} \frac{z^j q^{j(3j+1)/2}(-zq;q)_j}{(q;q)_j}.$$

Next, replace j by $j+1$ in the first sum and combine the two sums to discover

$$F(z) = (1+zq)F(zq). \tag{1.110}$$

Observe that $F(0) = 1$ and iterate (1.110) to obtain

$$F(z) = (-zq;q)_\infty.$$

$\square$

Cayley's technique of rewriting $1+zq^{2j}$ as $(1-q^j)+q^j(1+zq^j)$, splitting the sum into two pieces, and shifting the index of one of the sums, has often been used in the theory of q-series over the years in ever increasingly complicated situations.

1.11.3 Exercises

1. Fix a nonnegative integer j. Prove that

$$\sum_{n \geq 0} pd_j(n)q^n = \frac{q^{j(j+1)/2}}{(q;q)_j},$$

where $pd_j(n)$ denotes the number of partitions of n into distinct parts and where the number of parts is at most j.

2. Fix a nonnegative integer j. Prove that

$$\sum_{n=0}^{j(j+1)/2} \sum_{l=0}^{j} Q(j,l,n)z^l q^n = (-zq;q)_j,$$

where $Q(j,l,n)$ is the number of partitions of n into exactly l parts, all parts distinct and no greater than j.

3. Notice that the partition $\langle j^j \rangle$ is generated by $z^j q^{j^2}$.

4. Notice that in a partition $\lambda = (\lambda_1, \lambda_2, \ldots)$ of n that has Durfee square of order j, the jth largest part of λ, λ_j cannot be strictly less than j.

   (a) Use the above to find the two-variable generating function for the partitions λ into distinct parts with Durfee square of order j in which $\lambda_j > j$. (Hint: Think about such a λ in terms of a Ferrers graph decomposed into its Durfee square of order j, and a certain subpartition attached to the right of the Durfee square and another subpartition attached below the Durfee square.)

   (b) Use the above to find the two-variable generating function for the partitions λ into distinct parts with Durfee square of order j in which $\lambda_j = j$.

   (c) Sum the results in parts (a) and (b) to find the two-variable generating function for partitions into distinct parts with Durfee square of order j.

   (d) Sum your result from part (c) over all j to obtain a two-variable generating function for partitions into distinct parts.

   (e) Conclude the truth of Eq. (1.109). Compare this proof with Cayley's proof above.

Chapter 2

The Golden Age and its Modern Legacy

"It would be difficult to find more beautiful formulæ than the 'Rogers–Ramanujan' identities..."—G. H. Hardy [Ram27, p. xxxiv]

2.1 Hardy's Account

G. H. Hardy gave an account of the early history of the Rogers–Ramanujan identities in a lecture delivered at the Institute for Advanced Study in Princeton during the Fall 1936 semester. Fortunately, an account of this and the other lectures delivered by Hardy is preserved in the unpublished notes of Marshall Hall [Har37b]. An edited version of these comments, in which the reference to Ramanujan having found a proof for himself is omitted, appears in Hardy's later account [Har40, p. 90–91]. This later redacted account is the one that is often quoted, e.g. in [And76, p. 105] and [And82, pp. 1–2]. We shall reproduce the longer, more detailed account, from Hardy's IAS lecture here [Har37b, p. 28]:

> There are two theorems, the 'Rogers–Ramanujan identities', in which Ramanujan had been anticipated by a much less famous mathematician, but which are certainly two of the most remarkable formulæ which even he ever wrote down. The Rogers–Ramanujan identities are

$$1+\frac{q}{1-q}+\frac{q^4}{(1-q)(1-q^2)}+\cdots+\frac{q^{m^2}}{(1-q)(1-q^2)\cdots(1-q^m)}+\cdots$$
$$=\frac{1}{(1-q)(1-q^6)\cdots(1-q^4)(1-q^9)\cdots} \quad (2.1)$$

> and

$$1+\frac{q^2}{1-q}+\frac{q^6}{(1-q)(1-q^2)}+\cdots+\frac{q^{m(m+1)}}{(1-q)(1-q^2)\cdots(1-q^m)}+\cdots$$

$$= \frac{1}{(1-q^2)(1-q^7)\cdots(1-q^3)(1-q^8)\cdots}. \quad (2.2)$$

The exponents in the denominators on the right form in each case two arithmetical progressions with the difference 5. This is the surprise of the formulæ; the 'basic series' on the left are of a comparatively familiar type. The formulæ have a very curious history. They were found first, so long ago as 1894, by Rogers [L. J. Rogers, P.L.M.S. (1), 1894], a mathematician of great talent but comparatively little reputation, and one of whom very few people, had it not been for Ramanujan, might ever have heard. Rogers was a fine analyst, who anticipated 'Hölder's inequality' though without recognizing its importance or setting it in what is now its classical form. See Hardy, Littlewood, and Pólya, *Inequalities*, pp. 21–26. His gifts were, on a smaller scale, not unlike Ramanujan's own; but no one paid much attention to his work, and this particular paper was quite neglected.

Ramanujan rediscovered the formulæ some time before 1913, and stated them in the first of his letters to Hardy. He had then no proof (and knew that he had none); and neither Hardy nor MacMahon nor Perron could find one. They are therefore stated without proof in the second volume of MacMahon's *Combinatory Analysis*, published in 1916.

The mystery was solved, trebly, in 1917. In that year Ramanujan found a proof which will be given later. A little later he came accidentally across Rogers' paper and the more elaborate proof given there. Ramanujan was quite surprised by this find, and expressed the greatest admiration for Rogers' work. His rediscovery led incidentally to a belated recognition of Rogers' talent, and in particular to his election to the Royal Society. Finally I. Schur, who was then cut off from England by the war, rediscovered the identities again. Schur published two proofs [Berliner Sitzungsberichte, 1917, pp. 301–321] one of which is 'combinatorial' and quite unlike any other proof known. There are now seven published proofs, three by Rogers [one in the paper already referred to, one in Proc. Camb. Phil. Soc., 19, pp. 211–216, and one in P.L.M.S. (2), vol. 16 (1917), pp. 315–336, one by Ramanujan, two by Schur, and a later proof, based on quite different ideas, by Watson [J.L.M.S., vol. 4 (1929), pp. 4–9] and five at any rate of these proofs differ fundamentally. None of them is both simple and straightforward, and probably it would be unreasonable to hope for a proof which is. The simplest proofs are essentially verifications.

2.2 Two Proofs of the Rogers–Ramanujan Identities

Over the course of this book, we shall see many ways to prove the Rogers–Ramanujan identities.

2.2.1 Ramanujan's Proof

First we give the proof due to Ramanujan, alluded to by Hardy above. Recalling the notations of Definition 1.42 and Equation (1.59), the Rogers–Ramanujan identities may be stated as follows.

Theorem 2.1 (The Rogers–Ramanujan identities).

$$\sum_{n=0}^{\infty} \frac{q^{n^2}}{(q;q)_n} = \frac{f(-q^2, -q^3)}{(q;q)_\infty} \tag{2.3}$$

and

$$\sum_{n=0}^{\infty} \frac{q^{n(n+1)}}{(q;q)_n} = \frac{f(-q, -q^4)}{(q;q)_\infty}. \tag{2.4}$$

Proof. Define

$$F(z) := F(z;q) = \sum_{n=0}^{\infty} \frac{z^n q^{n^2}}{(q;q)_n}$$

and

$$G(z) := G(z;q) = \sum_{m=0}^{\infty} \frac{(-1)^m z^{2m} q^{m(5m-1)/2}(1 - zq^{2m})(z;q)_m}{(1-z)(q;q)_m}.$$

Notice that we may rewrite $G(z)$ as a bilateral sum,

$$G(z) = \sum_{m=-\infty}^{\infty} \frac{(-1)^m q^{m(5m-1)/2}(z;q)_m}{(1-z)(q;q)_m},$$

and we shall use the fact that by Jacobi's triple product identity (1.66), $G(1) = f(-q^2, -q^3)$ and $G(q) = f(-q, -q^4)$. Next, it is routine to verify that $G(z)$ satisfies the q-difference equation

$$G(z) = (1 - zq)G(zq) + zq(1 - zq)(1 - zq^2)G(zq^2), \tag{2.5}$$

or equivalently,

$$H(z) = H(zq) + zqH(zq^2), \tag{2.6}$$

where

$$H(z) := H(z,q) = \frac{G(z)}{(zq;q)_\infty}.$$

Also, it is easy to verify that

$$F(z) = F(zq) + zqF(zq^2),$$

and since $F(z)$ and $H(z)$ are power series satisfying the initial conditions $F(0) = H(0) = 1$, we may conclude

$$F(z) = H(z) = \frac{G(z)}{(zq;q)_\infty}. \tag{2.7}$$

Upon setting $z = 1$ in (2.7), we obtain (2.3), and upon setting $z = q$ in (2.7), we obtain (2.4). $\qquad\square$

And thus we have a "simple verification" proof of (2.3) and (2.4).

2.2.2 Watson's Proof

Watson's proof, according to Hardy, is "based on quite different ideas." Watson showed that the Rogers–Ramanujan identities may be deduced as a corollary of his q-analog of Whipple's theorem (Theorem 1.69).

Proof. Letting $b, c, d, e \to \infty$ in Equation (1.98) yields

$$_4\phi_7 \left[\begin{matrix} a, q\sqrt{a}, -q\sqrt{a}, q^{-n} \\ \sqrt{a}, -\sqrt{a}, aq^{n+1}, 0, 0, 0, 0 \end{matrix} ; q, a^2 q^{n+2} \right] = (aq;q)_n \, _1\phi_1 \left[\begin{matrix} q^{-n} \\ 0 \end{matrix} ; q, aq^{1+n} \right], \tag{2.8}$$

or

$$\sum_{k \geq 0} \frac{(1 - aq^{2k})(q^{-n};q)_k}{(1-a)(q;q)_k(aq^{n+1};q,k)} a^{2k} q^{(n+2)k+4\binom{k}{2}}$$

$$= (aq;q)_n \sum_{k \geq 0} \frac{(q^{-n};q)_k}{(q;q)_k}(-1)^k a^k q^{(n+1)k+\binom{k}{2}}. \tag{2.9}$$

Observe that

$$(q^{-n};q)_k q^{nk} = (q^n - 1)(q^n - q)(q^n - q^2) \cdots (q^n - q^{k-1}) \to (-1)^k q^{\binom{k}{2}}$$

as $n \to \infty$. Now let $n \to \infty$ in (2.9) to obtain

$$\sum_{k \geq 0} \frac{(1 - aq^{2k})}{(1-a)(q;q)_k} a^{2k} q^{k(5k-1)/2} = (aq;q)_\infty \sum_{k \geq 0} \frac{a^k q^{k^2}}{(q;q)_k}. \tag{2.10}$$

But the left side of (2.10) is $G(a,q)$, and the right side is $(a;q)_\infty F(a,q)$ from Ramanujan's proof. So the two Rogers–Ramanujan identities follow once again from setting $a = 1$ or q, and applying Jacobi's triple product identity (1.66) to the left side of Equation (2.10). $\qquad\square$

2.3   Rogers–Ramanujan–Bailey Machinery

2.3.1   Examples of Identities of Rogers–Ramanujan Type

The existence of the Rogers–Ramanujan identities is by no means an isolated phenomenon. There are many known identities of Rogers–Ramanujan type: Rogers included more than two dozen in his papers [Rog94, Rog17]; Ramanujan recorded an additional twenty in his lost notebook [AB05, AB09]; W. N. Bailey and F. J. Dyson included more than a dozen more in [Bai47, Bai48]; and Bailey's PhD student, L. J. Slater, produced a list of 130 identities of Rogers–Ramanujan type [Sla52], including most of the previously known results and many new ones. There were also some early contributions by F. H. Jackson [Jac28] and G. W. Starcher [Sta31], and much more recent contributions by I. Gessel and D. Stanton [GS83]; J. H. Loxton [Lox84]; McLaughlin and Sills (sometimes joint with Bowman or Zimmer) [BMS09, MS08, MS09, MSZ09]. The most comprehensive list of known Rogers–Ramanujan type identities in the literature is in the survey article by McLaughlin, Sills, and Zimmer [MSZ08]. See also Appendix A.

To give the reader a feel for the general shape of such identities, some examples are included below. While none of these *quite* rise to the level of austere beauty of (2.3) and (2.4), many arguably come close. In some cases, the identities have been named in the literature after the discoverer and the rediscoverer, as in the case of the Rogers–Ramanujan identities themselves.

There is a set of three "Rogers–Selberg identities," related to the modulus 7 [Rog94, Rog17, Sel36]; see also Dyson [Dys43]:

$$\sum_{n=0}^{\infty} \frac{q^{2n^2}}{(q^2;q^2)_n(-q;q)_{2n}} = \frac{f(-q^3,-q^4)}{(q^2;q^2)_\infty}, \tag{2.11}$$

$$\sum_{n=0}^{\infty} \frac{q^{2n(n+1)}}{(q^2;q^2)_n(-q;q)_{2n}} = \frac{f(-q^2,-q^5)}{(q^2;q^2)_\infty}, \tag{2.12}$$

$$\sum_{n=0}^{\infty} \frac{q^{2n(n+1)}}{(q^2;q^2)_n(-q;q)_{2n+1}} = \frac{f(-q,-q^6)}{(q^2;q^2)_\infty}. \tag{2.13}$$

The two "Jackson–Slater identities," related to the modulus 8 ([Jac28, p. 170]; [Sla52, pp. 11–12, Eqs. (39), (38)]) are

$$\sum_{n=0}^{\infty} \frac{q^{2n^2}}{(q;q)_{2n}} = \frac{f(q^3,q^5)}{(q^2;q^2)_\infty}, \tag{2.14}$$

$$\sum_{n=0}^{\infty} \frac{q^{2n(n+1)}}{(q;q)_{2n}} = \frac{f(q,q^7)}{(q^2;q^2)_\infty}. \tag{2.15}$$

As we shall see later, the following two Ramanujan–Slater identities [AB09, pp. 36–37, Entries 1.7.11–12], [Sla52, Eqs. (36) and (34)] are intimately linked with the Göllnitz–Gordon partition identities, Theorems 2.42 and 2.43:

$$\sum_{n=0}^{\infty} \frac{q^{n^2}(-q;q^2)_n}{(q^2;q^2)_n} = \frac{f(-q^3,-q^5)}{\psi(-q)}, \tag{2.16}$$

$$\sum_{n=0}^{\infty} \frac{q^{n(n+2)}(-q;q^2)_n}{(q^2;q^2)_n} = \frac{f(-q,-q^7)}{\psi(-q)}. \tag{2.17}$$

There are three identities due to Bailey related to the modulus 9 [Bai47, p. 422, Eqs. (1.7)–(1.9)]:

$$\sum_{n=0}^{\infty} \frac{q^{3n^2}(q;q)_{3n}}{(q^3;q^3)_n(q^3;q^3)_{2n}} = \frac{f(-q^4,-q^5)}{(q^3;q^3)_\infty}, \tag{2.18}$$

$$\sum_{n=0}^{\infty} \frac{q^{3n(n+1)}(q;q)_{3n}(1-q^{3n+2})}{(q^3;q^3)_n(q^3;q^3)_{2n+1}} = \frac{f(-q^4,-q^5)}{(q^3;q^3)_\infty}, \tag{2.19}$$

$$\sum_{n=0}^{\infty} \frac{q^{3n(n+1)}(q;q)_{3n+1}}{(q^3;q^3)_n(q^3;q^3)_{2n+1}} = \frac{f(-q,-q^7)}{(q^3;q^3)_\infty}. \tag{2.20}$$

There are three identities due to Rogers related to the modulus 14 [Rog94, p. 341, Ex. 2], [Rog17, p. 329 (1)]:

$$\sum_{n=0}^{\infty} \frac{q^{n^2}}{(q;q^2)_n(q;q)_n} = \frac{f(-q^6,-q^8)}{(q;q)_\infty}, \tag{2.21}$$

$$\sum_{n=0}^{\infty} \frac{q^{n(n+1)}}{(q;q^2)_{n+1}(q;q)_n} = \frac{f(-q^4,-q^{10})}{(q;q)_\infty}, \tag{2.22}$$

$$\sum_{n=0}^{\infty} \frac{q^{n(n+2)}}{(q;q^2)_{n+1}(q;q)_n} = \frac{f(-q^2,-q^{12})}{(q;q)_\infty}. \tag{2.23}$$

The four identities of Dyson related to the modulus 27 [Bai47, p. 12, Eqs. (B1)–(B4)] are as follows:

$$\sum_{n=0}^{\infty} \frac{q^{n(n+3)}(q^3;q^3)_n}{(q;q)_{2n+2}(q;q)_n} = \frac{f(-q^3,-q^{24})}{(q;q)_\infty}, \tag{2.24}$$

$$\sum_{n=0}^{\infty} \frac{q^{n(n+2)}(q^3;q^3)_n}{(q;q)_{2n+2}(q;q)_n} = \frac{f(-q^6,-q^{21})}{(q;q)_\infty}, \tag{2.25}$$

$$\sum_{n=0}^{\infty} \frac{q^{n(n+1)}(q^3;q^3)_n}{(q;q)_{2n+1}(q;q)_n} = \frac{f(-q^9,-q^{18})}{(q;q)_\infty}, \tag{2.26}$$

$$1 + \sum_{n=1}^{\infty} \frac{q^{n^2}(q^3;q^3)_{n-1}}{(q;q)_{2n-1}(q;q)_n} = \frac{f(-q^{12},-q^{15})}{(q;q)_\infty}. \tag{2.27}$$

Eq. (2.26), in which the right-hand side simplifies to $\prod_{m\geq 1}(1-q^{9m})/(1-q^m)$, is Dyson's personal favorite. On June 2, 1987, at the Ramanujan Centenary Conference, Dyson shared, "I found a lot of identities of the sort that Ramanujan would have enjoyed. My favorite was... [Eq. (2.26)]. In the cold dark evenings, while I was scribbling these beautiful identities amid the death and destruction of 1944, I felt close to Ramanujan. He had been scribbling even more beautiful identities amid the death and destruction of 1917." [Dys88]

2.3.2 Bailey's Transform and Bailey Pairs

During the winter of 1943–1944, Bailey undertook a study of certain aspects of Rogers' work on the Rogers–Ramanujan identities and identities of similar type, and managed to simplify and expand upon Rogers' work [Bai47, Bai48]. According to Bailey, he began this course of research without having seen Rogers' papers [Rog93a, Rog94, Rog95, Rog17][1] for some years. To the best of his knowledge, the only copies in Manchester, where he lived and worked, had been destroyed in the blitz of 1940.

Hardy, as editor of the *Proceedings of the London Mathematical Society*, enlisted Dyson (who was about 20 years old at the time, and had taken classes with Hardy at Cambridge) to referee Bailey's submission. Hardy, realizing that Bailey and Dyson were perhaps the only two people in England at the time with a serious interest in the Rogers–Ramanujan identities, dispensed with the usual custom of keeping the referee anonymous, and a correspondence between Dyson and Bailey ensued. Dyson and Bailey never met face to face, but Dyson saved the letters he received from Bailey. A transcription of these letters is included as an appendix to this volume. Due to paper shortages caused by the war, Bailey's papers did not appear until 1947 and 1948.

Theorem 2.2 (Bailey's transform). *If*

$$\beta_n = \sum_{r=0}^{n} \alpha_r u_{n-r} v_{n+r} \tag{2.28}$$

and

$$\gamma_n = \sum_{r=n}^{\infty} \delta_r u_{r-n} v_{r+n},$$

then

$$\sum_{n=0}^{\infty} \alpha_n \gamma_n = \sum_{n=0}^{\infty} \beta_n \delta_n.$$

---

[1]R. Askey recommends reading the papers [Rog93a, Rog94, Rog95] in reverse order: start with the third memoir on the expansion of certain infinite products [Rog95], proceed to the second [Rog94], and finish with the first [Rog93a].

Proof.

$$\sum_{n=0}^{\infty} \alpha_n \gamma_n = \sum_{n=0}^{\infty}\sum_{r=n}^{\infty} \alpha_n \delta_r u_{r-n} v_{r+n} = \sum_{r=0}^{\infty}\sum_{n=0}^{r} \delta_r \alpha_n u_{r-n} v_{r+n} = \sum_{r=0}^{\infty} \delta_r \beta_r.$$

$\square$

Remark 2.3. Bailey, as an analyst, inserts the caveat: "We assume, of course, that the series converge and that the change in the order of summation is allowable. In all the cases given in this paper the justification would be simple." [Bai48, p. 1]

Definition 2.4. A pair of sequences $\big(\alpha_n(x,q), \beta_n(x,q)\big)$ that satisfies (2.28) with $u_n = 1/(q;q)_n$ and $v_n = 1/(xq;q)_n$, i.e.,

$$\beta_n = \sum_{r=0}^{n} \frac{\alpha_r}{(q;q)_{n-r}(xq;q)_{n+r}}, \tag{2.29}$$

is called a *Bailey pair with respect to x.*

Remark 2.5. In order to apply standard q-hypergeometric theorems, it will often be convenient to rewrite the Bailey pair relationship (2.29) in the form

$$\beta_n = \frac{1}{(q;q)_n(xq;q)_n} \sum_{r=0}^{n} \frac{(q^{-n};q)_r}{(xq^{n+1};q)_r}(-1)^r q^{nr-\binom{r}{2}}\alpha_r. \tag{2.30}$$

Note also that Equation (2.30) can be inverted:

$$\alpha_n = (1 - xq^{2n}) \sum_{r=0}^{n} \frac{(-1)^{n-r} q^{\binom{n-r}{2}}(xq;q)_{n+r-1}}{(q;q)_{n-r}}\beta_r. \tag{2.31}$$

Lemma 2.6 (Bailey's Lemma). *If $(\alpha_n(x,q), \beta_n(x,q))$ form a Bailey pair, then so do $(\alpha_n'(x,q), \beta_n'(x,q))$, where*

$$\alpha_r'(x,q) = \frac{(\rho_1;q)_r(\rho_2;q)_r}{(xq/\rho_1;q)_r(xq/\rho_2;q)_r}\left(\frac{xq}{\rho_1\rho_2}\right)^r \alpha_r(x,q) \tag{2.32}$$

and

$$\beta_n'(x,q)$$
$$= \frac{1}{(xq/\rho_1;q)(xq/\rho_2;q)_n} \sum_{j=0}^{n} \frac{(\rho_1;q)_j(\rho_2;q)_j(xq/\rho_1\rho_2)_{n-j}}{(q;q)_{n-j}}\left(\frac{xq}{\rho_1\rho_2}\right)^j \beta_j(x,q). \tag{2.33}$$

Equivalently, if (α_n, β_n) *form a Bailey pair, then*

$$\frac{1}{(xq/\rho_1; q)_n (xq/\rho_2; q)_n} \sum_{j=0}^n \frac{(\rho_1; q)_j (\rho_2; q)_j (xq/\rho_1 \rho_2; q)_{n-j}}{(q; q)_{n-j}} \left(\frac{xq}{\rho_1 \rho_2} \right)^j \beta_j$$

$$= \sum_{r=0}^n \frac{(\rho_1; q)_r (\rho_2; q)_r}{(q; q)_{n-r} (xq; q)_{n+r} (xq/\rho_1; q)_r (xq/\rho_2; q)_r} \left(\frac{xq}{\rho_1 \rho_2} \right)^r \alpha_r. \quad (2.34)$$

Next, we include a corollary of Bailey's lemma, which can be used to obtain two-variable generalizations of Rogers–Ramanujan type identities.

Corollary 2.7.

$$\sum_{n=0}^\infty x^n q^{n^2} \beta_n(x, q) = \frac{1}{(xq; q)_\infty} \sum_{r=0}^\infty x^r q^{r^2} \alpha_r(x, q) \quad (2.35)$$

$$\sum_{n=0}^\infty x^n q^{n^2} (-q; q^2)_n \beta_n(x, q^2) = \frac{(-xq; q^2)_\infty}{(xq^2; q^2)_\infty} \sum_{r=0}^\infty \frac{x^r q^{r^2} (-q; q^2)_r}{(-xq; q^2)_r} \alpha_r(x, q^2)$$

$$\quad (2.36)$$

$$\sum_{n=0}^\infty x^n q^{n(n+1)/2} (-1; q)_n \beta_n(x, q) = \frac{(-xq; q)_\infty}{(xq; q)_\infty} \sum_{r=0}^\infty \frac{x^r q^{r(r+1)/2} (-1; q)_r}{(-xq; q)_r} \alpha_r(x, q)$$

$$\quad (2.37)$$

Proof. Let $n, \rho_2 \to \infty$ in (2.34) to obtain

$$\sum_{j=0}^\infty \frac{(-1)^j x^j q^{j(j+1)/2} (\rho_1; q)_j}{\rho_1^j} \beta_j(x, q)$$

$$= \frac{(xq/\rho_1; q)_\infty}{(xq; q)_\infty} \sum_{r=0}^\infty \frac{(-1)^r x^r q^{r(r+1)/2} (\rho_1; q)_r}{\rho_1^r (xq/\rho_1; q)_r} \alpha_r(x, q). \quad (2.38)$$

To obtain (2.36), let $\rho_1 \to \infty$ in (2.38). To obtain (2.36), in (2.38), set $\rho_1 = -\sqrt{q}$ and replace q by q^2 throughout. To obtain (2.37), in (2.38) set $\rho_1 = -q$. □

The following easy corollary can be used to obtain a plethora of Rogers–Ramanujan type identities immediately from Bailey pairs.

Corollary 2.8.

$$\sum_{n=0}^\infty q^{n^2} \beta_n(1, q) = \frac{1}{(q; q)_\infty} \sum_{r=0}^\infty q^{r^2} \alpha_r(1, q) \quad (2.39)$$

$$\sum_{n=0}^\infty q^{n^2} (-q; q^2)_n \beta_n(1, q^2) = \frac{1}{\psi(-q)} \sum_{r=0}^\infty q^{r^2} \alpha_r(1, q^2) \quad (2.40)$$

$$\sum_{n=0}^{\infty} q^{n(n+1)/2}(-1;q)_n \beta_n(1,q) = \frac{2}{\varphi(-q)} \sum_{r=0}^{\infty} \frac{q^{r(r+1)/2}}{1+q^r} \alpha_r(1,q) \quad (2.41)$$

$$\frac{1}{1-q} \sum_{n=0}^{\infty} q^{n(n+1)/2}(-q;q)_n \beta_n(q,q) = \frac{1}{\varphi(-q)} \sum_{r=0}^{\infty} q^{r(r+1)/2} \alpha_r(q,q) \quad (2.42)$$

Proof. To obtain (2.39), (2.40), and (2.41), set $x = 1$ in (2.35), (2.36), and (2.37) respectively. To obtain (2.42), set $x = q$ and $\rho_2 = -q$ in (2.38). $\square$

Remark 2.9. The "Bailey transform" was so named by Slater [Sla66]. The appellations "Bailey pair" and "Bailey's lemma" are due to Andrews.

2.3.3 A General Bailey Pair and Its Consequences

As we shall see, all of the identities due to Rogers and Bailey, and a number of additional identities due to Slater and others, may be derived as special cases of a single, rather general Bailey pair and an associated set of q-difference equations.

Definition 2.10. For positive integers d, e, and k, let

$$\alpha_n^{(d,e,k)}(x^e, q^e) := \frac{(-1)^{n/d} x^{(k/d-1)n} q^{(k-d+1/2)n^2/d - n/2}(1-xq^{2n})}{(1-x)(q^d;q^d)_{n/d}}$$
$$\times (x;q^d)_{n/d} [\![d \mid n]\!], \quad (2.43)$$

where

$$[\![P(n,d)]\!] = \begin{cases} 1 & \text{if } P(n,d) \text{ is true}, \\ 0 & \text{if } P(n,d) \text{ is false} \end{cases}$$

is the Iverson bracket, and let

$$\beta_n^{(d,e,k)}(x^e, q^e) = \frac{1}{(q^e;q^e)_n (x^e q^e;q^e)_n}$$
$$\times \sum_{r=0}^{\lfloor n/d \rfloor} \frac{(x;q^d)_r (1-xq^{2dr})(q^{-en};q^e)_{dr}}{(q^d;q^d)_r (1-x)(x^e q^{e(n+1)};q^e)_{dr}} (-1)^{(d+1)r} x^{(k-d)r} q^{Q(d,e,k;n,r)},$$
$$(2.44)$$

where $Q(d, e, k; n, r) = (2k - 2d - ed + 1)\frac{d}{2}r^2 + (e-1)\frac{d}{2}r + endr$.

Remark 2.11. The Iverson bracket (or "truth function") was introduced by Kenneth E. Iverson in his 1962 book on the APL computer language [Ive62]. Donald Knuth advocated for the use of this notation in [Knu92] hoping that it would soon gain universal acceptance. This has not happened, however, as some authors prefer to use the notation $\chi(P)$ over $[\![P]\!]$.

Proposition 2.12. $\left(\alpha_n^{(d,e,k)}(x,q), \beta_n^{(d,e,k)}(x,q)\right)$ *form a Bailey pair.*

Proof. This is a direct application of Eq. (2.30). □

In some sense, Proposition 2.12, by itself, is vacuous; it is true by definition. The *real point* is as follows: $\alpha_n^{(d,e,k)}(x,q)$ is deliberately reverse-engineered so that upon inserting it into a limiting case of Bailey's lemma with $x = 1$ (Corollary 2.8), Jacobi's triple product identity can be invoked to yield an infinite product. Further, we wish to choose specific values of d, e, and k so that (2.44) can be subjected to a q-hypergeometric transformation or summation formula yielding an expression that is "nice" in some sense. So let us proceed.

Definition 2.13.

$$H_{k,i}(a;x;q) := \frac{(axq;q)_\infty}{(x;q)_\infty} \sum_{n=0}^\infty \frac{x^{kn}q^{kn^2+(1-i)n}a^n(1/a;q)_n(1-x^iq^{2ni})(x;q)_n}{(q;q)_n(axq;q)_n}.$$

Remark 2.14. Notice that the $a = 0$ case is well defined since

$$\lim_{a\to 0} a^n(1/a;q)_n = \lim_{a\to 0} \prod_{j=0}^{n-1}(a-q^j) = (-1)^n q^{n(n-1)/2}.$$

Theorem 2.15. *For positive integers d, e, and k,*

$$\sum_{n=0}^\infty x^{en}q^{en^2}\beta_n^{(d,e,k)}(x^e,q^e) = \frac{(xq^d;q^d)_\infty}{(x^eq^e;q^e)_\infty} H_{d(e-1)+k,1}(0;x;q^d).$$

Proof. Although the algebraic details may appear forbidding at first, the result follows directly from the definitions. □

Let us consider $\beta^{(1,1,2)}(x,q)$. We wish to transform it from the form given in the definition to a simpler form.

$$\beta_n^{(1,1,2)}(x,q) = \frac{1}{(q;q)_n(xq;q)_n} \sum_{r=0}^n \frac{(x;q)_r(1-xq^{2r})(q^{-n};q)_r}{(q;q)_r(1-x)(xq^{n+1})_r} x^r q^{r^2+nr}$$

$$= \frac{1}{(q;q)_n(xq;q)_n} \lim_{b\to\infty} {}_6\phi_5 \left[\begin{matrix} x, q\sqrt{x}, -q\sqrt{x}, q^{-n}, b, b \\ \sqrt{x}, -\sqrt{x}, xq^{n+1}, xq/b, xq/b \end{matrix} ; q, \frac{xq^{n+1}}{b^2} \right]$$

$$= \frac{1}{(q;q)_n(xq;q)_n} \lim_{b\to\infty} \frac{(xq;q)_n(xq/b^2;q)_n}{(xq/b;q)_n^2} \quad \text{(by Theorem 1.50)}$$

$$= \frac{1}{(q;q)_n}.$$

Thus the $(d,e,k) = (1,1,2)$ case of Theorem 2.15 is

$$\sum_{n=0}^{\infty} \frac{x^n q^{n^2}}{(q;q)_n} = H_{2,1}(0;x;q) = \frac{1}{(x;q)_\infty} \sum_{n=0}^{\infty} \frac{(-1)^n x^{2n} q^{\frac{5}{2}n^2 - \frac{1}{2}n}(1 - xq^{2n})(x;q)_n}{(q;q)_n}.$$

(2.45)

Upon setting $x = 1$ and applying Jacobi's triple product identity to the right-hand side, (or equivalently, by setting $x = 1$ first, and inserting the resulting Bailey pair into (2.39)) we obtain the first Rogers–Ramanujan identity, Eq. (2.3),

$$\sum_{n=0}^{\infty} \frac{q^{n^2}}{(q;q)_n} = \frac{f(-q^2, -q^3)}{(q;q)_\infty}.$$

But wait! There is more! Having established the Bailey pair

$$\left(\alpha_n^{(1,1,2)}(x;q), \beta_n^{(1,1,2)}(x;q)\right) = \left(\frac{1}{(q;q)_n}, \frac{(-1)^n x^n q^{\frac{3}{2}n^2 - \frac{1}{2}n}(1 - xq^{2n})(x;q)_n}{(1-x)(q;q)_n}\right)$$

for the purpose of deriving the first Rogers–Ramanujan identity, we may quickly derive two additional identities "for free." By inserting the same Bailey pair (with $x = 1$) into (2.40) and (2.41), we obtain, respectively, the first Ramanujan–Slater mod 8 identity (which is associated with the first Göllnitz–Gordon partition theorem), listed earlier as Eq. (2.16),

$$\sum_{n=0}^{\infty} \frac{q^{n^2}(-q;q^2)_n}{(q^2;q^2)_n} = \frac{f(-q^3, -q^5)}{\psi(-q)},$$

and a special case of Lebesgue's identity (Theorem 1.74), which Ramanujan recorded in the Lost Notebook [AB09, p. 38 , Entry 1.7.14],

$$\sum_{n=0}^{\infty} \frac{q^{n(n+1)/2}(-1;q)_n}{(q;q)_n} = \frac{f(-q^2, -q^2)}{\varphi(-q)}.$$

(2.46)

Let us now derive a simple form for $\beta_n^{(1,2,2)}(x,q)$.

$$\beta_n^{(1,2,2)}(x^2, q^2) = \frac{1}{(q^2;q^2)_n(x^2q^2;q^2)_n}$$

$$\times \sum_{r=0}^{n} \frac{(x;q)_r(1 - xq^{2r})(q^{-2n};q^2)_r}{(q;q)_r(1-x)(x^2q^{2n+2};q^2)_r} x^r q^{\frac{1}{2}r^2 + \frac{1}{2}r + 2nr}$$

$$= \frac{1}{(q^2;q^2)_n(x^2q^2;q^2)_n}$$

$$\times \lim_{b \to \infty} {}_6\phi_5 \left[\begin{matrix} x, q\sqrt{x}, -q\sqrt{x}, q^{-n}, -q^{-n}, b \\ \sqrt{x}, -\sqrt{x}, xq^{n+1}, -xq^{n+1}, xq/b \end{matrix} ; q, -\frac{xq^{2n+1}}{b} \right]$$

$$= \frac{1}{(q^2;q^2)_n(x^2q^2;q^2)_n} \lim_{b \to \infty} \frac{(xq;q)_n(-xq^{n+1}/b;q)_n}{(xq/b;q)_n(-xq^{n+1};q)_n}$$

$$(\text{by Theorem } 1.50)$$

$$= \frac{1}{(q^2;q^2)_n(-xq;q)_{2n}}.$$

Thus the $(d,e,k) = (1,2,2)$ case of Equation (2.15) is

$$\sum_{n=0}^{\infty} \frac{x^{2n}q^{2n^2}}{(q^2;q^2)_n(-xq;q)_{2n}} = \frac{(xq;q)_\infty}{(x^2q^2;q^2)_\infty}H_{3,1}(0;x;q), \qquad (2.47)$$

which in the $x = 1$ case is the first Rogers–Selberg identity (2.11).

We may also insert the $(1,2,2)$-Bailey pair with $x = 1$ into Eq. (2.40) to obtain another identity of Rogers related to the modulus 5 [Rog94, p. 330],

$$\sum_{n=0}^{\infty} \frac{q^{n^2}}{(q^4;q^4)_n} = \frac{f(-q,-q^4)}{\psi(-q)}. \qquad (2.48)$$

Remark 2.16. The first Bailey mod 9 identity, Equation (2.18), arises from the case $(d,e,k) = (1,3,2)$. The first Rogers mod 14 identity, Equation (2.21), arises from the case $(2,1,3)$. The first Dyson mod 27 identity, Equation (2.27), arises from $(3,1,4)$.

The question naturally arises: for *which* values of (d,e,k) may we reasonably expect to find β_n representable as a finite product? The key, in the two cases we have considered, was the application of Theorem 1.50, Jackson's ${}_6\phi_5$ summation formula, in order to sum the expression for the β_n. To achieve this, the numerator and denominator parameters in the q-hypergeometric representation of $\beta_n^{(d,e,k)}(x^e,q^e)$ must be limited to 6 and 5, respectively. To this end, we note that

$$(q^{-en};q^e)_{dr} = \prod_{j=0}^{d-1}(q^{-en+ej};q^{de})_r, \qquad (2.49)$$

and that each factor on the right-hand side of (2.49) can be factored into a product of e factors as

$$(q^{en+ej};q^{de})_r = \prod_{h=1}^{e}(\xi_e^h q^{j-n};q^d)_r,$$

where ξ_e is a primitive eth root of unity. (The complementary denominator factor $(x^e q^{e(n+1)};q^e)_{dr}$ can be analogously expressed as a product of ed rising q-factorials. Additionally, the factor $q^{(2k-2d-ed+1)\frac{d}{2}r^2}$ can be written as a limit of a product of $|2k - ed - 2d + 1|$ rising q-factorials.)

Thus, $\beta_n^{(d,e,k)}(x^e,q^e)$ can be expressed as a finite product times a limiting case of a very-well-poised ${}_{m+1}\phi_m$, where

$$m = ed + |2k - ed - 2d + 1| + 2.$$

Thus, the positive integer solutions (d, e, k) to

$$ed + |2k - ed - 2d + 1| + 2 \leq 5 \tag{2.50}$$

give rise to Rogers–Ramanujan type identities of the form "infinite product equals single-fold sum."

These solutions are $(1, 1, 1), (1, 1, 2), (1, 2, 1), (1, 2, 2), (2, 1, 2), (2, 1, 3)$, and $(3, 1, 4)$.

Remark 2.17. Interest in Rogers–Ramanujan type identities involving a double series seems to have begun with Andrews [And75] and increased with Verma and Jain [VJ80, VJ82]. Additional identities of this type were given by S. O. Warnaar [War01a] and A. V. Sills [Sil04a, Sil05, Sil06, Sil07]. Indeed, values of (d, e, k) for which $ed + |2k - ed - 2d + 1| + 2 = 7$ will naturally lead to the expression for $\beta_n^{(d,e,k)}(x^e, q^e)$ to involve a very-well-poised $_8\phi_7$, which cannot be summed, but can be transformed via Watson's q-Whipple theorem (Theorem 1.69) into a finite product times a balanced $_4\phi_3$. The substitution of the resulting form into limiting cases of Bailey's lemma yields double series-product identities.

2.3.4 q-Difference Equations

The alert reader will have noticed that we have only derived via Bailey pairs the *first* of the two Rogers–Ramanujan identities, the *first* of the three Rogers–Selberg identities, etc. While it is certainly possible to derive all members of a given family via insertion of an appropriate Bailey pair into an instance of Bailey's lemma (as in Slater's papers [Sla51b, Sla52]), we prefer to utilize systems of q-difference equations associated with a given Bailey pair to obtain complete families of identities. To prepare, we will require some definitions and results first established by Andrews [And68].

Remark 2.18. The notation we are about to introduce is standard throughout the literature. As a result, first and second Rogers–Ramanujan identities are associated with the cases $i = 2$ and 1, respectively, the opposite of what we might expect. Although surprising and potentially confusing to the uninitiated, it is actually part of a general scheme which will eventually seem more natural.

Definition 2.19.

$$\begin{aligned}
J_{k,i}(a; x; q) &:= H_{k,i}(a; xq; q) - axq H_{k,i-1}(a; xq; q) \\
&= \frac{(axq; q)_\infty}{(xq; q)_\infty} \sum_{n=0}^{\infty} \frac{a^n x^{kn} q^{kn^2 + (k-i+1)n} (1/a; q)_n (xq; q)_n}{(q; q)_n (axq; q)_{n+1}} \\
&\quad \times \left(1 - axq^{n+1} + ax^i q^{2ni+i-n} - x^i q^{i+2ni}\right).
\end{aligned}$$

Lemma 2.20.

$$H_{k,i}(a; x; q) - H_{k,i-1}(a; x; q) = x^{i-1} J_{k,k-i+1}(a; x; q).$$

Proof.

$$H_{k,i}(a;x;q) - H_{k-1}(a;x;q)$$

$$= \sum_{n=0}^{\infty} \frac{x^{kn}q^{kn^2+n-in}a^n(axq^{n+1};q)_\infty(1/a;q)_n(1-q^n)}{(q;q)_n(xq^n;q)_\infty}$$

$$+ \sum_{n=0}^{\infty} \frac{x^{kn+i-1}q^{kn^2+n+n(i-1)}a^n(axq^{n+1};q)_\infty(1/a;q)_n(1-xq^n)}{(q;q)_n(xq^n;q)_\infty}$$

$$= \sum_{n=1}^{\infty} \frac{x^{kn}q^{kn^2+n-in}a^n(axq^{n+1};q)_\infty(1/a;q)_n}{(q;q)_{n-1}(xq^n;q)_\infty}$$

$$+ \sum_{n=0}^{\infty} \frac{x^{kn+i-1}q^{kn^2+n+n(i-1)}a^n(axq^{n+1};q)_\infty(1/a;q)_n}{(q;q)_n(xq^n;q)_\infty}$$

$$= \sum_{n=0}^{\infty} \frac{x^{kn+k}q^{kn^2+n+2kn+k=1-in-i}a^{n+1}(axq^{n+2};q)_\infty(1/a;q)_{n+1}}{(q;q)_n(xq^{n+1};q)_\infty}$$

$$+ \sum_{n=0}^{\infty} \frac{x^{kn+i-1}q^{kn^2+n+n(i-1)}a^n(axq^{n+1};q)_\infty(1/a;q)_n}{(q;q)_n(xq^n;q)_\infty}$$

$$= x^{i-1}\sum_{n=0}^{\infty} \frac{x^{kn}q^{kn^2+in}a^n(axq^{n+2};q)_\infty(1/a;q)_n}{(q;q)_n(xq^{n+1};q)_\infty}$$

$$\times \left((1-axq^{n+1}) + ax^{k-i+1}q^{2n(k-1)+k-i+1+n}(1-a^{-1}q^n)\right)$$

$$= x^{i-1}\sum_{n=0}^{\infty} \frac{x^{kn}q^{kn^2+in}a^n(axq^{n+2};q)_\infty(1/a;q)_n}{(q;q)_n(xq^{n+1};q)_\infty} \left(1-(xq)^{k-i+1}q^{2n(k-i+1)}\right)$$

$$+ x^{i-1}\sum_{n=0}^{\infty} \frac{x^{kn}q^{kn^2+in}a^n(axq^{n+2};q)_\infty(1/a;q)_n}{(q;q)_n(xq^{n+1};q)_\infty}$$

$$\times \left(axq^{n+1}\left(1-(xq)^{k-i}q^{2n(k-i)}\right)\right)$$

$$= x^{i-1}\left(H_{k,k-i+1}(a;xq;q) - axqH_{k,k-i}(a;xq;q)\right)$$

$$= x^{i-1}J_{k,k-i+1}(a;x;q),$$

where we have strategically used Cayley's observation that

$$q^{-in}(1-x^iq^{2ni}) - q^{-n(i-1)}(1-x^{i-1}q^{2n(i-1)})$$
$$= q^{-in}(1-q^n) + x^{i-1}q^{n(n-1)}(1-xq^n).$$

$\square$

Lemma 2.21.

$$J_{k,i}(a;x;q) = J_{k,i-1}(a;x;q) + (xq)^{i-1}\left(J_{k,k-i+1}(a;xq;q) - aJ_{k,k-i+2}(a;xq;q)\right) \tag{2.51}$$

Proof.

$$J_{k,i}(a;x;q) - J_{k,i-1}(a;x;q)$$
$$= H_{k,i}(a;xq;q) - H_{k,i-1}(a;xq;q) - axqH_{k,i-1}(a;xq;q) + axqH_{k,i-2}(a;xq;q)$$
$$= (xq)^{i-1}\big(J_{k,k-i+1}(a;xq;q) - aJ_{k,k-i+2}(a;xq;q)\big).$$

□

Lemma 2.22.

$$H_{k,1}(a;x;q) = J_{k,k}(a;x;q) = J_{k,k+1}(a;x;q) \tag{2.52}$$
$$H_{k,0}(a;x;q) = 0 \tag{2.53}$$
$$H_{k,-i}(a;x;q) = -x^i H_{k,i}(a;x;q) \tag{2.54}$$

Proof. Exercise.

□

Definition 2.23. For positive integers d, e, and k,

$$Q_i^{(d,e,k)}(x) := \frac{(xq^d;q^d)_\infty}{(x^e q^e;q^e)_\infty} J_{d(e-1)+k,i}(0;x;q^d).$$

Theorem 2.24. *For* $1 \leq i \leq k + d(e-1)$,

$$Q_i^{(d,e,k)}(x) = Q_{i-1}^{(d,e,k)}(x) + \frac{(1 - xq^d)x^{i-1}q^{d(i-1)}}{(x^e q^e;q^e)_d} Q_{d(e-1)+k-i+1}^{(d,e,k)}(xq^d) \tag{2.55}$$

and $Q_j^{(d,e,k)}(0) = 1$, *for* $1 \leq j \leq d(e-1) + k$.

Lemma 2.25. *For* $i \leq i \leq k$,

$$J_{k,i}(0;1;q) = \frac{f(-q^i, -q^{2k-i+1})}{(q;q)_\infty}.$$

Proof.

$$J_{k,i}(0;1;q) = H_{k,i}(0;q;q)$$
$$= \frac{1}{(q;q)_\infty}\sum_{n=0}^{\infty}(-1)^n q^{(2k+1)n(n+1)/2 - in}(1 - q^{(2n+1)i})$$
$$= \frac{f(-q^i, -q^{2k-i+1})}{(q;q)_\infty} \qquad \text{(by Eq. (1.66))}.$$

□

Corollary 2.26.

$$Q_i^{(d,e,k)}(1) = \frac{f(-q^{di}, -q^{d(2ed-2d+2k+1)})}{f(-q^e)}.$$

Proof.

$$Q_i^{(d,e,k)}(1) = \frac{(q^d; q^d)_\infty}{(q^e; q^e)_\infty} J_{d(e-1)+k,i}(0; 1; q^d)$$

$$= \frac{(q^d; q^d)_\infty}{(q^e; q^e)_\infty} \frac{f(-q^{di}, -q^{d(2ed-2d+2k+1)})}{f(-q^d)} \quad \text{(by Lemma 2.25)}.$$

$\square$

Definition 2.27. For positive integers d, e, and k, let

$$F_{d(e-1)+k}^{(d,e,k)} := \sum_{n=0}^\infty x^{en} q^{en^2} \beta_n^{(d,e,k)}(x^e, q^e).$$

Theorem 2.28. *For positive integers d, e, and k,*

$$F_{d(e-1)+k}^{(d,e,k)}(x) = Q_{d(e-1)+k}^{(d,e,k)}(x).$$

Proof.

$$F_{d(e-1)+k}^{(d,e,k)}(x) = \frac{(xq^d; q^d)_\infty}{(x^e q^e; q^e)_\infty} H_{d(e-1)+k,1}(0; x; q^d)$$

$$= \frac{(xq^d; q^d)_\infty}{(x^e q^e; q^e)_\infty} J_{d(e-1)+k,d(e-1)+k}(0; x; q^d)$$

$$= Q_{d(e-1)+k}^{(d,e,k)}(x).$$

$\square$

Once we have $F_{d(e-1)+k}^{(d,e,k)}(x)$ for a given (d, e, k), we can then use the system of q-difference equations to derive $F_i^{(d,e,k)}(x)$ for $1 \le i \le d(e-1) + k - 1$.

Example 2.29. The case $(d, e, k) = (2, 1, 3)$. We already have

$$F_3^{(2,1,3)}(x) = \sum_{n=0}^\infty \frac{x^n q^{n^2}}{(xq; q^2)_n (q; q)_n}, \tag{2.56}$$

and the system of q-difference equations

$$F_1^{(2,1,3)}(x) = \frac{1}{1 - xq} F_3^{(2,1,3)}(xq^2) \tag{2.57}$$

$$F_2^{(2,1,3)}(x) = F_1^{(2,1,3)}(x) + \frac{xq^2}{1 - xq} F_2^{(2,1,3)}(xq^2) \tag{2.58}$$

$$F_3^{(2,1,3)}(x) = F_2^{(2,1,3)}(x) + \frac{x^2 q^4}{1 - xq} F_1^{(2,1,3)}(xq^2). \tag{2.59}$$

From (2.57) and (2.56), we find

$$F_1^{(2,1,3)}(x) = \frac{1}{1-xq} F_3^{(2,1,3)}(xq^2) = \sum_{n=0}^{\infty} \frac{x^n q^{n^2+2n}}{(xq;q^2)_{n+1}(q;q)_n}. \tag{2.60}$$

Then, turning around Eq. (2.59), and for convenience defining

$$\phi(x) := \sum_{n=0}^{\infty} \frac{x^{n+1} q^{n^2+3n+1}}{(xq;q^2)_{n+1}(q;q)_n}, \tag{2.61}$$

we find

$$F_2^{(2,1,3)}(x) = F_3^{(2,1,3)}(x) - \frac{x^2 q^4}{1-xq} F_1^{(2,1,3)}(xq^2)$$

$$= F_3^{(2,1,3)}(x) - \sum_{n=0}^{\infty} \frac{x^{n+2} q^{(n+2)^2}}{(xq;q^2)_{n+2}(q;q)_n}$$

$$= F_3^{(2,1,3)}(x) - \sum_{n=1}^{\infty} \frac{x^{n+1} q^{(n+1)^2}(1-q^n)}{(xq;q^2)_{n+1}(q;q)_n}$$

$$= F_3^{(2,1,3)}(x) - \sum_{n=0}^{\infty} \frac{x^{n+1} q^{n^2+2n+1}}{(xq;q^2)_{n+1}(q;q)_n} + \phi(x)$$

$$= F_3^{(2,1,3)}(x) - \sum_{n=0}^{\infty} \frac{x^n q^{n^2}(1-q^n)(1-xq^{2n+1})}{(xq;q^2)_{n+1}(q;q)_n} + \phi(x)$$

$$= F_3^{(2,1,3)}(x) - \sum_{n=0}^{\infty} \frac{x^n q^{n^2}(1-xq^{2n+1}-q^n+xq^{3n+1})}{(xq;q^2)_{n+1}(q;q)_n} + \phi(x)$$

$$= F_3^{(2,1,3)}(x) - F_3^{(2,1,3)}(x) + \sum_{n=0}^{\infty} \frac{x^n q^{n^2+n}}{(xq;q^2)_n(q;q)_n} - \phi(x) + \phi(x)$$

$$= \sum_{n=0}^{\infty} \frac{x^n q^{n^2+n}}{(xq;q^2)_n(q;q)_n}.$$

Thus $F_i^{(2,1,3)}(1) = Q_i^{(2,1,3)}(1)$ for $i = 1,2,3$ correspond respectively to Equations (2.23), (2.22), and (2.21).

Remark 2.30. For additional discussion of Bailey's transform, lemma, and pairs, see Andrews' q-series monograph [And86] and the survey article of McLaughlin, Sills, and Zimmer [MSZ08]. For additional parametrized Bailey pairs and their associated q-difference equations, see [Sil04a, Sil05, Sil06, Sil07].

2.3.5 Another Bailey Pair and a Pair of Two-Variable Identities of Ramanujan

We next derive a simple, yet powerful Bailey pair. Its proof does not require the power of the full hypergeometric machinery, but rather only the finite q-binomial theorem.

Lemma 2.31. *The pair* (A_n, B_n), *where*

$$A_n = A_n(1, q) = \begin{cases} (a^n + a^{-n})q^{n^2/2} & \text{if } n > 0 \\ 1 & \text{if } n = 0. \end{cases}$$

and

$$B_n = B_n(1, q) = \frac{(-\sqrt{q}/a; q)_n(-a\sqrt{q}; q)_n}{(q; q)_{2n}},$$

forms a Bailey pair with respect to 1.

Proof.

$$B_n = \frac{1}{(q;q)_n^2} + \sum_{r=1}^{n} \frac{(a^r + a^{-r})q^{r^2/2}}{(q;q)_{n-r}(q;q)_{n+r}}$$

$$= \sum_{r=-n}^{n} \frac{a^r q^{r^2/2}}{(q;q)_{n-r}(q;q)_{n+r}}$$

$$= \frac{1}{(q;q)_{2n}} \sum_{r=-n}^{n} \binom{2n}{n+r}_q a^r q^{r^2/2}$$

$$= \frac{1}{(q;q)_{2n}} \sum_{s=0}^{2n} \binom{2n}{s}_q a^{s-n} q^{(s-n)^2/2}$$

$$= \frac{a^{-n} q^{n^2/2}}{(q;q)_{2n}} (-aq^{(1-2n)/2}; q)_{2n} \quad \text{(by Equation (1.56))}$$

$$= \frac{(-aq^{1/2}; q)_n(-a^{-1}q^{1/2}; q)_n}{(q;q)_{2n}}.$$

$\square$

The novelty of the Bailey pair (A_n, B_n) is that upon inserting into (2.39) or (2.40), we may derive series–product identities in *both* variables q and a:

Theorem 2.32 (Ramanujan).

$$\sum_{n=0}^{\infty} \frac{q^{2n^2}(-q/a; q^2)_n(-aq; q^2)_n}{(q^2; q^2)_{2n}} = \frac{f(aq^3, q^3/a)}{(q^2; q^2)_\infty} \tag{2.62}$$

$$\sum_{n=0}^{\infty} \frac{q^{n^2}(-q/a; q^2)_n(-qa; q^2)_n}{(q; q^2)_n(q^4; q^4)_n} = \frac{f(aq^2, q^2/a)}{\psi(-q)} \tag{2.63}$$

Many Rogers–Ramanujan type identities follow from Theorem 2.32 by setting a equal to ρq^r for a complex root of unity ρ and rational number r. For a further discussion and generalizations of Theorem 2.32, see the paper by McLaughlin and Sills [MS12]. Similarly, Andrews' q-analogue of Gauss's $_2F_1(1/2)$ sum (Theorem 1.76) and Andrews' q-Bailey sum (Theorem 1.77) contain many Rogers–Ramanujan type identities as special cases.

2.3.6   Exercises

1. Prove that (2.29), (2.30), and (2.31) are in fact equivalent.

2. Derive the following forms:

   (a)
   $$\beta_n^{(1,1,1)}(x,q) = \delta_{n,0} := \begin{cases} 1 & \text{if } n = 0 \\ 0 & \text{otherwise} \end{cases} .$$

   (The function $\delta_{n,0}$ is known as the *Kronecker delta function*.) Note that
   $$\left(\frac{(-1)^n q^{n(n-1)/2}(1 - xq^{2n})(x;q)_n}{(1-x)(q;q)_n}, \delta_{n,0} \right)$$
   is known as the *unit Bailey pair*.

   (b) $\beta_n^{(1,2,1)}(x^2,q^2) = \dfrac{(-1)^n q^{n^2}}{(q^2;q^2)_n(-xq;q)_{2n}}.$

   (c) $\beta_n^{(1,3,2)}(x^3,q^3) = \dfrac{(xq;q)_{3n}}{(q^3;q^3)_n(x^3q^3;q^3)_{2n}}.$

   (d) $\beta_n^{(2,1,2)}(x,q) = \dfrac{q^{n(n-1)/2}}{(q;q)_n(xq;q^2)_n}.$

   (e) $\beta_n^{(2,1,3)}(x,q) = \dfrac{1}{(q;q)_n(xq;q^2)_n}.$

   (f) $\beta_n^{(3,1,4)}(x,q) = \dfrac{(x;q^3)_n}{(q;q)_n(x;q)_{2n}}.$

3. Using, in turn, each of the Bailey pairs associated with the β_n in Exercise 2, derive each of the following Rogers–Ramanujan type identities:

   (a) $1 = 1$.

   (b) $\displaystyle\sum_{n=0}^{\infty} \frac{(-1)^n q^{3n^2}}{(-q;q^2)_n(q^4;q^4)_n} = \frac{f(-q^2,-q^3)}{(q^2;q^2)_\infty}$ [Rog94, p. 339, Ex. 2], cf. [Sla52, p. 154, Eq. (19)].

   (c) Equation (2.18).

   (d) $\displaystyle\sum_{n=0}^{\infty} \frac{q^{n(3n-1)/2}}{(q;q^2)_n(q;q)_n} = \frac{f(-q^4,-q^6)}{(q;q)_\infty}$ [Rog94, p. 341, Ex. 1], cf. [Sla52, p. 156, Eq. (46)].

(e) Equation (2.21).

(f) Equation (2.27).

4. Using, in turn, the associated system of q-difference equations, derive the following "siblings" to the identities in Exercise 2.

(a) (None.)

(b) $$\sum_{n=0}^{\infty} \frac{(-1)^n q^{n(3n-2)}}{(-q;q^2)_n (q^4;q^4)_n} = \frac{f(-q,-q^4)}{(q^2;q^2)_\infty}$$ [Rog17, p. 330, Eq. (5)],

cf. [Sla52, p. 154, Eq. (15)], and $$\sum_{n=0}^{\infty} \frac{(-1)^n q^{n(3n+2)}}{(-q;q^2)_{n+1}(q^4;q^4)_n} = \frac{f(-q,-q^4)}{(q^2;q^2)_\infty}$$

[AB05, p. 252, Eq. (11.2.7)].

(c) Equations (2.19), and (2.20). Why are there only three such Bailey mod 9 identities instead of four?

(d) $$\sum_{n=0}^{\infty} \frac{q^{3n(n+1)/2}}{(q;q^2)_{n+1}(q;q)_n} = \frac{f(-q^2,-q^8)}{(q;q)_\infty}$$ [Rog17, p. 330, Eq. (2), line 2],

cf. [Sla52, p. 156, Eq. (44)], and $$\sum_{n=0}^{\infty} \frac{q^{n(3n+1)/2}}{(q;q^2)_n (q;q)_n} = \frac{f(-q^4,-q^6)}{(q;q)_\infty}$$

[Rog17, p. 330 (2), line 2].

(e) Equations (2.22) and (2.23).

(f) Equations (2.26), (2.25), and (2.24).

5. (a) Prove that
$$\left(q^{-\binom{n}{2}} x^{-n} \alpha_n^{(1,1,2)}(x,q), \frac{1}{(q^2;q^2)_n} \right)$$
is a Bailey pair with respect to x.

(b) Use this Bailey pair to derive the identities

$$\sum_{n=0}^{\infty} \frac{q^{n^2}}{(q^2;q^2)_n} = \frac{\psi(-q)}{(q;q)_\infty} \tag{2.64}$$

and

$$\sum_{n=0}^{\infty} \frac{q^{n(n+1)}}{(q^2;q^2)_n} = \frac{\varphi(-q^2)}{(q;q)_\infty}. \tag{2.65}$$

6. Derive the following identities:

(a)
$$\sum_{n=0}^{\infty} \frac{q^{3n^2}(q^2;q^2)_{3n}}{(q^{12};q^{12})_n (q^3;q^3)_{2n}} = \frac{f(-q^5,-q^7)}{\psi(-q^3)}$$
(Dyson [Bai48, p. 9, Eq. (7.5)])

(b)

$$\sum_{n=0}^{\infty} \frac{q^{4n^2}(q;q^2)_{2n}}{(q^4;q^4)_{2n}} = \frac{f(-q,-q^{11})}{(q^4;q^4)_\infty}$$

(Slater [Sla52, Eq. (53)])

(c)

$$\sum_{n=0}^{\infty} \frac{q^{2n^2}(-q^{-1};q^2)_n(-q^3;q^2)_n}{(q^2;q^2)_{2n}} = \frac{f(q,q^5)}{(q^2;q^2)_\infty}$$

(Stanton [Sta01])

(d)

$$\sum_{n=0}^{\infty} \frac{q^{2n^2}(-q;q^2)_{2n}}{(q^8;q^8)_n(q^2;q^4)_n} = \frac{f(q^3,q^5)}{\psi(-q^2)}$$

(Gessel and Stanton [GS83, Eq. (7.24)])

(e)

$$\sum_{n=0}^{\infty} \frac{q^{5n^2}(q^2;q^5)_n(q^3;q^5)_n}{(q^5;q^5)_{2n}} = \frac{f(-q^7,-q^8)}{(q^5;q^5)_\infty}$$

7. (a) Prove that

$$\left((a^{n+1}+a^{-n})q^{\binom{n+2}{2}}, \frac{(-a;q)_{n+1}(-q/a;q)_n(1-q)}{(q;q)_{2n+1}}\right)$$

is a Bailey pair with respect to q.

(b) Insert the Bailey pair of part (a) into (2.42) and derive a two-variable series-product identity.

8. Verify Eqs. (2.52), (2.53), and (2.54).

2.4   Combinatorial Considerations

With regard to the Rogers–Ramanujan identities (2.3) and (2.4), Hardy reports [Har37b, p. 31], "neither Rogers nor Ramanujan considered their combinatorial aspect." (The same can be said of Bailey and Slater.) The first to realize their combinatorial significance were P. A. MacMahon [Mac18, p. 33] and I. Schur [Sch17]. As Hardy had noted [Har37b], the German Schur, while cut off from England by World War I, had independently rediscovered the Rogers–Ramanujan identities (including their combinatorial interpretation). Around the same time, Ramanujan showed the identities to MacMahon (although Ramanujan had yet to find a proof, and was not yet aware of Rogers' original discovery of them). Thus in [Mac18, p. 33], MacMahon amusingly writes, "This most remarkable theorem has been verified as far as the coefficient of x^{89} by actual expansion so that there is practically no reason to doubt its truth; but it has not yet been established," and then proceeds to explain the partition theoretic interpretation.

2.4.1 Combinatorial Rogers–Ramanujan Identities

Theorem 2.33 (First Rogers–Ramanujan identity: combinatorial version).
The number of partitions of n into parts which mutually differ from each other by at least 2 equals the number of partitions of n into parts $\equiv \pm 1$ (mod 5).

Proof. Let $r_1(n)$ denote the number of partitions of n into parts which mutually differ from each other by at least 2, and let $r_1^*(n)$ denote the number of partitions of n into parts congruent to ± 1 (mod 5). The proof consists of demonstrating that the left and right-hand sides of (2.3) are the generating functions of $r_1(n)$ and $r_1^*(n)$ respectively.

We first consider the right-hand side of (2.3),

$$\frac{f(-q^2,-q^3)}{(q;q)_\infty} = \frac{(q^2,q^3,q^5;q^5)_\infty}{(q;q)_\infty} = \frac{1}{(q;q^5)_\infty(q^4;q^5)_\infty},$$

and thus, by reasoning due to Euler (see Section 1.1.1),

$$\sum_{n=0}^{\infty} r_1^*(n)q^n = \frac{1}{(q;q^5)_\infty(q^4;q^5)_\infty}.$$

Next, recall that the generating function for $p_l(n)$, the number of partitions into at most l parts is

$$\frac{1}{(q;q)_l}.$$

A given partition $\lambda = (\lambda_1,\lambda_2,\ldots,\lambda_m)$ of length m where $m \leq l$, may be identified with the l-tuple $(\lambda_1,\lambda_2,\ldots,\lambda_m,0,0,\ldots,0)$ where $\lambda_1 \geq \lambda_2 \geq \cdots \geq \lambda_m$ and where the number of trailing zeros is $l-m$. (In other words, a partition with at most l parts may be padded on the right with zeros, to create an l-tuple.) Thus the generating function for the number of such l-tuples of weakly decreasing nonnegative integers is also

$$\frac{1}{(q;q)_l}.$$

To this l-tuple, increase the first component by $2l-1$, the second by $2l-3$, ..., the penultimate by 3, and the last by 1, to obtain

$$(\lambda_1+2l-1,\lambda_2+2l-3,\ldots,\lambda_2+3,\lambda_1+1). \tag{2.66}$$

Thus, the generating function for l-tuples of the type (2.66) is

$$\frac{q^{(2l-1)+(2l-3)+\cdots+5+3+1}}{(q;q)_l} = \frac{q^{l^2}}{(q;q)_l}.$$

Note that (2.66) is a partition of length l of the type enumerated by $r_1(n)$. Thus by summing over all possible lengths l, we find

$$\sum_{l=0}^{\infty} \frac{q^{l^2}}{(q;q)_l} = \sum_{n=0}^{\infty} r_1(n)q^n.$$

And so, by (2.3), $r_1(n) = r_1^*(n)$ for all $n \geq 0$. $\square$

Theorem 2.34 (Second Rogers–Ramanujan identity: combinatorial version). *The number of partitions of n into parts greater than 1 which mutually differ from each other by at least 2 equals the number of partitions of n into parts $\equiv \pm 2 \pmod 5$.*

Proof. The proof is analogous to that of Theorem 2.33, and is thus left as an exercise. $\square$

There is a natural combinatorial interpretation to the "x-generalizations" of the Rogers–Ramanujan identities $F_2^{(1,1,2)}(x)$ and $F_1^{(1,1,2)}(x)$ as well as their associated q-difference equations. We have

$$F_1^{(1,1,2)}(x) = \sum_{l=0}^{\infty} \frac{x^l q^{l^2+l}}{(q;q)_l} = \sum_{n=0}^{\infty}\sum_{l=0}^{n} r_2(l,n)x^l q^n,$$

$$F_2^{(1,1,2)}(x) = \sum_{l=0}^{\infty} \frac{x^l q^{l^2}}{(q;q)_l} = \sum_{n=0}^{\infty}\sum_{l=0}^{n} r_1(l,n)x^l q^n,$$

where $r_j(x,q)$ denotes the number of partitions of n of length l into parts greater than $j-1$ which mutually differ by at least 2. From Theorem 2.24, the associated system of q-difference equations is

$$F_1^{(1,1,2)}(x) = F_2^{(1,1,2)}(xq) \tag{2.67}$$

$$F_2^{(1,1,2)}(x) = F_1^{(1,1,2)}(x) + xq F_1^{(1,1,2)}(xq). \tag{2.68}$$

Note that replacing x by xq in one of these two-variable generating functions amounts to increasing each part by 1 in every partition generated. A moment's reflection reveals that 2.67 is telling us that if we start with the set of partitions into parts which mutually differ by at least 2, and increase each part by 1, the set of partitions obtained will be those in which parts differ by at least 2 and each part is greater than 1.

Let us rewrite (2.68) as

$$F_2^{(1,1,2)}(x) - F_1^{(1,1,2)}(x) = xq F_2^{(1,1,2)}(xq). \tag{2.69}$$

The left-hand side of (2.69) generates partitions into parts differing by at least 2 in which the smallest part is exactly 1, by subtracting partitions enumerated by $r_2(n)$ from those enumerated by $r_1(n)$. The right-hand side of (2.69) generates those same partitions by starting with partitions into parts greater than

1 and parts differing by at least 2, increasing each part by 1 and appending a
1. The increase of each part by 1 is a result of having $F_2(xq)$ instead of $F_2(x)$,
and the appending of a 1 is a result of multiplying by xq.

Historical Note 2.35. George Andrews shared the following (G. E. Andrews, e-mail message to the author, September 28, 2009):

> [L. J.] Mordell was here [visiting Penn State] more than once in the
> late '60s as a guest of S. Chowla. ...He was always interested in
> the work of young mathematicians. "What are you working on?"
> was his question to me.
>
> "Partitions!" I replied.
>
> "I never liked partitions," he responded rather coldly. "What exactly is it that you do with them?"
>
> I said that I had been studying the Rogers–Ramanujan identities.
>
> "Ahh!" he replied much more warmly. "But those are *analytic*
> identities!"

2.4.2 Schur's Partition Identity Related to the Modulus 6

The next Rogers–Ramanujan type partition theorem is due to Schur in
1926 [Sch26].

Theorem 2.36. *Let $s(n)$ denote the number of partitions λ of n in which*

$$\lambda_i - \lambda_{i+1} \geq 3 \tag{2.70}$$

and in which

$$\lambda_i - \lambda_{i+1} = 3 \implies 3 \nmid \lambda_i, \tag{2.71}$$

*for $i = 1, 2, \ldots, l(\lambda) - 1$; i.e. $s(n)$ is the number of partitions of n into parts
that mutually differ by at least 3 and in which no consecutive multiples of 3
appear. Let $s^*(n)$ denote the number of partitions of n into parts congruent to
$\pm 1 \pmod 6$. Then $s(n) = s^*(n)$ for all integers n.*

There are many known proofs of Schur's partition identity. Particularly
noteworthy is David Bressoud's bijective proof [Bre80b] (cf. [And86, pp. 56–
58]).

However, here we choose to present a proof that emphasizes the connection between Schur's identity and a q-analog of the trinomial theorem (see
Section 1.8.2). This proof is due to Andrews [And94].

Define

$$S_n(x, q) := \sum_{h \geq 0} \sum_{j \geq 0} s_n(h, j) x^h q^j, \tag{2.72}$$

where $s_n(h, j)$ is the number of partitions λ of j satisfying conditions (2.70)
and (2.71), and the additional conditions

$$\lambda_1 \leq n \tag{2.73}$$

and

$$\#\{\lambda_i : \lambda_i \equiv 1 \,(\mathrm{mod}\ 3)\} - \#\{\lambda_i : \lambda_i \equiv 2 \,(\mathrm{mod}\ 3)\} = h. \tag{2.74}$$

Lemma 2.37.

$$S_{3n-1}(x,q) = \sum_{j=-n}^{n} x^j q^{j(3j-1)/2} \left(\binom{n}{j} \right)_q^3. \tag{2.75}$$

Proof. Let us begin with an examination of the generating function (2.72) and consider the cases where the largest allowable part lies in each of the three possible residue classes modulo 3, i.e., consider $S_{3n}(x,q)$, $S_{3n-1}(x,q)$ and $S_{3n-2}(x,q)$. For each of these three cases, we introduce a difference equation by breaking the generating function into a sum of two generating functions the first, which enumerates the partitions where the largest possible part does not in fact appear, and the second, in which it does. Thus we find:

$$S_{3n}(x,q) = S_{3n-1}(x,q) + q^{3n} S_{3n-4}(x,q) \tag{2.76}$$

$$S_{3n-1}(x,q) = S_{3n-2}(x,q) + x^{-1} q^{3n-1} S_{3n-4}(x,q) \tag{2.77}$$

$$S_{3n-2}(x,q) = S_{3n-3}(x,q) + x q^{3n-2} S_{3n-5}(x,q) \tag{2.78}$$

By strategic substitution and solving among the first order recurrences in the above system, we may obtain the following second order recurrence:

$$S_{3n-1}(x,q) = (1 + xq^{3n-2} + x^{-1} q^{3n-1}) S_{3n-4}(x,q) + q^{3n-3}(1 - q^{3n-3}) S_{3n-7}(x,q). \tag{2.79}$$

Further, note the initial condition $S_{-1}(x,q) = 1$ and $S_2(x,q) = 1 + xq + x^{-1} q^2$. But these are the initial conditions and recurrence satisfied by $\mathscr{T}_n(x,q)$, in the context of a q-analog of the trinomial theorem (1.90), and thus

$$S_{3n-1}(x,q) = \mathscr{T}_n(x,q)$$

for all $n \geq 0$. □

With Lemma 2.37 in hand, we may now prove Schur's theorem.

Proof of Theorem 2.36.

$$\sum_{m=0}^{\infty} s(m) q^m = \lim_{n \to \infty} S_{3n-1}(1,q)$$

$$= \lim_{n \to \infty} \sum_{j=-n}^{n} q^{j(3j-1)/2} \left(\binom{n}{j} \right)_q^3 \quad \text{(by Lemma 2.37)}$$

$$= \frac{1}{(q^3; q^3)_\infty} \sum_{j=-\infty}^{\infty} q^{j(3j-1)/2} \quad \text{(by Theorem 1.67)}$$

$$= \frac{f(q, q^2)}{(q^3; q^3)_\infty} \quad \text{(by Jacobi's triple product identity (1.66))}$$

$$= (-q, q^3)_\infty (-q^2; q^3)_\infty$$

$$= \frac{(q^2; q^6)_\infty (q^4; q^6)_\infty}{(q; q^3)_\infty (q^2; q^3)_\infty}$$

$$= \frac{1}{(q; q^6)_\infty (q^5; q^6)_\infty}$$

$$= \sum_{m=0}^{\infty} s^*(m) q^m$$

□

Remark 2.38. Recall Theorem 1.9 to conclude that $s(n)$ also equals the number of partitions into distinct nonmultiples of 3 and $s(n)$ equals the number of partitions of n into odd parts which may appear at most twice each.

Remark 2.39. Andrews, Katherin Bringmann, and Karl Mahlburg [ABM15] present an analytic identity that represents Schur's partition theorem:

$$\sum_{n=0}^{\infty} \sum_{m=0}^{\infty} \frac{(-1)^n q^{3n(3n+2m)+m(3m-1)/2}}{(q; q)_m (q^6; q^6)_n} = \frac{1}{(q; q^6)_\infty (q^5; q^6)_\infty}.$$

2.4.3 A False Lead

Let $Q_d(n)$ denote the number of partitions of n into parts that mutually differ by at least d.

Let $Q_d^*(n)$ denote the number of partitions of n into parts which mutually differ by at least d and in which no consecutive multiples of d appear.

We may rephrase Euler's partition theorem (Theorem 1.8) as

The number of partitions of n into parts $\equiv \pm 1 \pmod 4$ equals $Q_1(n)$.

Similarly, the combinatorial version of the first Rogers–Ramanujan identity (Theorem 2.33) states

The number of partitions of n into parts $\equiv \pm 1 \pmod 5$ equals $Q_2(n)$.

Schur's theorem (Theorem 2.36) states

The number of partitions of n into parts $\equiv \pm 1 \pmod 6$ equals $Q_3^*(n)$.

It is natural to wonder whether there are analogous partition identities involving $Q_d(n)$ or $Q_d^*(n)$ and partitions with parts in some arithmetic sequences related to the modulus $d + 3$, for $d > 3$.

The answer is *no!*

In the 1940s, Derrick Lehmer and Henry Alder closed that door definitively. Lehmer proved [Leh46, p. 541, Theorem 2] the following theorem.

Theorem 2.40 (Lehmer's nonexistence theorem). *The number $Q_d(n)$ of partitions into parts differing by at least d is NOT equal to the number of partitions of n taken from any set S of integers whatsoever, except when $d = 0$, 1, or 2.*

Alder [Ald48, p. 713, Theorem 3] proved the following companion to Lehmer's theorem.

Theorem 2.41 (Alder's nonexistence theorem). *The number $Q_d^*(n)$ of partitions into parts differing by at least d and in which no consecutive multiples of d occur is NOT equal to the number of partitions of n taken from any set S of integers whatsoever if $d > 3$.*

The theorems of Lehmer and Alder had a chilling effect on the field. In reality, all that they did was to prove that the particular sequence of results suggested at the beginning of this subsection did not continue, or as George Andrews so aptly put it: "The wrong generalization is the wrong generalization." Nonetheless, mathematicians in the 1950s acted as though "the Rogers–Ramanujan identities do not generalize," (which, as we shall later see, is patently false) and ceased looking for possible generalizations. (We will not interrupt the narrative to provide proofs of these nonexistence theorems here; the interested reader may consult the indicated references.)

Also, the following obvious question left open by Alder's theorem was not asked. Noting that Alder's theorem proves nonexistence in the cases $d > 3$, and that the $d = 1$ and 3 cases correspond to the first Rogers–Ramanujan identity and Schur's theorem respectively, what about the case $d = 2$?

The world would have to wait until the 1960s for two brilliant gentlemen, one a German undergraduate and one a young American professor, to independently discover that something very interesting indeed happens in the $d = 2$ case.

2.4.4   Göllnitz–Gordon Identities

The following two partition theorems were first discovered by Heinz Göllnitz as a part of his baccalaureate thesis in 1960 [Göl60], although he did not publish them until 1967 [Göl67]. The identities were independently rediscovered by Basil Gordon, a professor at UCLA, who published them in 1965 [Gor65]. (Gordon had completed his PhD at Caltech in 1956 under the direction of Tom Apostol, who in turn had been a student of D. H. Lehmer at Berkeley. Thus the chilling of the field initiated by Lehmer and his student Alder began to be thawed by Lehmer's academic grandchild, Basil Gordon.)

Theorem 2.42 (First Göllnitz–Gordon identity). *Let $G_1(n)$ denote the number of partitions of n into parts that mutually differ by at least 2 and in which no consecutive even numbers appear as parts. Let $G_1^*(n)$ denote the number of partitions of n into parts congruent to $\pm 1, 4 \pmod 8$. Then $G_1(n) = G_1^*(n)$ for all n.*

Theorem 2.43 (Second Göllnitz–Gordon identity). *Let $G_3(n)$ denote the number of partitions of n into parts greater than 2 that mutually differ by at least 2 and in which no consecutive even numbers appear. Let $G_3^*(n)$ denote* *the number of partitions of n into parts congruent to* 3, 4, 5 (mod 8). *Then* $G_3(n) = G_3^*(n)$ *for all* n.

One is immediately struck by the similarities between the Rogers–Ramanujan identities and Schur's partition theorem.

Proof of Theorem 2.42. Recall the first Ramanujan–Slater identity:

$$\sum_{m=0}^{\infty} \frac{q^{m^2}(-q;q^2)_m}{(q^2;q^2)_m} = \frac{f(-q^3,-q^5)}{\psi(-q)}. \tag{2.16}$$

Observe that

$$
\begin{aligned}
\frac{f(-q^3,-q^5)}{\psi(-q)} &= \frac{f(-q^3,-q^5)}{f(-q,-q^3)} \\
&= \frac{(q^3;q^8)_\infty(q^5;q^8)_\infty(q^8;q^8)_\infty}{(q;q^4)_\infty(q^3;q^4)_\infty(q^4;q^4)_\infty} \\
&= \frac{(q^3;q^8)_\infty(q^5;q^8)_\infty(q^8;q^8)_\infty}{(q;q^8)_\infty(q^5;q^8)_\infty(q^3;q^8)_\infty(q^7;q^8)_\infty(q^4;q^8)_\infty(q^8;q^8)_\infty} \\
&= \frac{1}{(q;q^8)_\infty(q^4;q^8)_\infty(q^7;q^8)_\infty} \\
&= \sum_{n=0}^{\infty} G_1^*(n)q^n.
\end{aligned}
$$

Next consider the following refinement of the left-hand side of (2.16):

$$
\begin{aligned}
\sum_{m=0}^{\infty} & x_1^{2m-1}x_2^{2m-3}\cdots x_m \\
& \times \frac{(1+y_1)(1+y_1^2 y_2)(1+y_1^2 y_2^2 y_3)\cdots(1+y_1^2 y_2^2 y_3^2\cdots y_{m-1}^2 y_m)}{(1-z_1^2)(1-(z_1 z_2)^2)(1-(z_1 z_2 z_3)^2)\cdots(1-(z_1 z_2\cdots z_m)^2)}. \tag{2.80}
\end{aligned}
$$

Clearly, if all x_i, y_i, and z_i are set equal to q, we will recover the left side of (2.16). By including subscripts and three different variable names, we will be able to track the parts of every partition generated. Notice that the mth term of (2.80) generates a triple of partitions (α, β, γ) as follows: The factor $x_1^{2m-1}x_2^{2m-3}\cdots x_{m-1}^3 x_m$ generates the partition $\alpha = (2m-1, 2m-3, \ldots, 3, 1)$. The factors

$$(1+y_1)(1+y_1^2 y_2)(1+y_1^2 y_2^2 y_3)\cdots(1+y_1^2 y_2^2 y_3^2\cdots y_{m-1}^2 y_m)$$

generate partitions β which can most conveniently be described as linear combinations of partitions[2]:

$$\beta = b_1(1) + b_2(2,1) + b_3(2,2,1) + b_4(2,2,2,1) + \cdots + b_m(2,2,2,\ldots,2,1),$$

---

[2]Sums of partitions were defined in Definition 1.4 and scalar multiples of partitions have the naïve meaning $b(\lambda_1, \lambda_2, \ldots, \lambda_m) = (b\lambda_1, \ldots, b\lambda_m)$.

where each $b_i = 0$ or 1.

The factors

$$\frac{1}{(1 - z_1^2)(1 - (z_1 z_2)^2)(1 - (z_1 z_2 z_3)^2) \cdots (1 - (z_1 z_2 \cdots z_m)^2)}$$

generate partitions γ of length at most m in which all parts are even, thus $\gamma = (2c_1, 2c_2, \ldots, 2c_m)$ for some integers $c_1 \geq c_2 \geq \cdots \geq c_m \geq 0$.

Thus it remains to show that there is a bijection between those partitions of length m that are enumerated by $G_1(n)$, and the triple (α, β, γ) of partitions described above.

First we observe that $\lambda = \alpha + \beta + \gamma$ is a partition of length m in which $\lambda_i - \lambda_{i+1} \geq 2$ for $i = 1, 2, \ldots, m-1$ because of the form of α. Next, we observe that the minimum difference between parts λ_i and λ_{i+1} can only occur when $\beta_i = \beta_i = 2$, and $\gamma_i = \gamma_{i+1}$. Recalling that $\alpha_i - \alpha_{i+1} = 2$, we observe that in this case

$$\lambda_i - \lambda_{i+1} = \alpha_{i+1} - \alpha_i + \beta_i - \beta_{i+1} + \gamma_i - \gamma_{i+1}$$
$$= 2 + 0 + 0,$$

but this forces λ_i and λ_{i+1} to both be odd by the parities of α_i, α_{i+1} (which are always odd), β_i, β_{i+1} (which are both 2 in this minimal difference case), and γ_i, γ_{i+1} (which are always even). Thus λ is a partition enumerated by $G_1(n)$.

The other direction of the bijection requires us to start with a partition λ of length m, difference of at least 2 between parts, and no consecutive even integers appearing as parts, and decompose it into a sum of partitions $\alpha + \beta + \gamma$ with the forms of α, β, and γ as defined above.

Given λ of length m, α is given as $(2m - 1, 2m - 3, \ldots, 3, 1)$. Form

$$\beta = b_1(1) + b_2(2, 1) + b_3(2, 2, 1) + \cdots + b_m(2, 2, 2, \ldots, 2, 1)$$
$$= \Big(b_1 + 2(b_2 + b_3 + \cdots + b_m), b_2 + 2(b_3 + b_4 + \cdots + b_m),$$
$$\ldots, b_{m-1} + 2b_m, b_m\Big)$$

where

$$b_i = \begin{cases} 1 & \text{if } \lambda_i \text{ is even,} \\ 0 & \text{if } \lambda_i \text{ is odd} \end{cases}.$$

Thus

$$\beta_i \equiv b_i \equiv 1 - \lambda_i \pmod{2}. \tag{2.81}$$

All that remains is to show that $\lambda - \alpha - \beta$ is a partition of length at most m in which all parts are even, and thus is of the required form for γ.

Using (2.81), we obtain

$$\gamma_i := \lambda_i - \alpha_i - \beta_i \equiv \lambda_i - 1 - (1 - \lambda_i) \equiv 0 \pmod{2},$$

i.e. all parts of γ are even.

Noting $\beta_i - \beta_{i+1} = b_i + b_{i+1}$, let us consider the possible cases: If λ_i and λ_{i+1} are both even,

$$\begin{aligned} \gamma_i - \gamma_{i+1} &= (\lambda_i - \lambda_{i+1}) - (\alpha_i - \alpha_{i+1}) - (\beta_i - \beta_{i+1}) \\ &\geq 4 - 2 - 2 \\ &= 0. \end{aligned}$$

If λ_i and λ_{i+1} are both odd,

$$\begin{aligned} \gamma_i - \gamma_{i+1} &= (\lambda_i - \lambda_{i+1}) - (\alpha_i - \alpha_{i+1}) - (\beta_i - \beta_{i+1}) \\ &\geq 2 - 2 - 0 \\ &= 0. \end{aligned}$$

If one of λ_i and λ_{i+1} is even and the other is odd,

$$\begin{aligned} \gamma_i - \gamma_{i+1} &= (\lambda_i - \lambda_{i+1}) - (\alpha_i - \alpha_{i+1}) - (\beta_i - \beta_{i+1}) \\ &\geq 3 - 2 - 1 \\ &= 0. \end{aligned}$$

Thus $\lambda_i - \lambda_{i+1}$ is always nonnegative, and so λ is a partition.

$\square$

The proof of Theorem 2.43 is similar and will therefore be omitted.

2.4.5 Little Göllnitz Partition Identities

Closely related to the Göllnitz–Gordon identities are the "little" Göllnitz identities (so named by K. Alladi [All98, p. 155], to distinguish them from the "big" Göllnitz partition identity, to be discussed later).

Theorem 2.44 (First little Göllnitz identity). *Let $LG_1(n)$ denote the number of partitions of n into parts that mutually differ by at least 2 and in which no consecutive odd numbers appear as parts. Let $LG_1^*(n)$ denote the number of partitions of n into distinct parts not congruent to 3 modulo 4. Let $LG_1^{**}(n)$ denote the number of partitions of n into parts congruent to $1, 5,$ or 6 (mod 8). Then $LG_1(n) = LG_1^*(n) = LG_1^{**}(n)$ for all n.*

Theorem 2.45 (Second little Göllnitz identity). *Let $LG_2(n)$ denote the number of partitions of n into parts that mutually differ by at least 2 and in which no consecutive odd numbers appear as parts, and no part equals 1. Let $LG_2^*(n)$ denote the number of partitions of n into distinct parts not congruent to 1 modulo 4. Let $LG_2^{**}(n)$ denote the number of partitions of n into parts congruent to $2, 3,$ or 7 (mod 8). Then $LG_2(n) = LG_2^*(n) = LG_2^{**}(n)$ for all n.*

Remark 2.46. Notice that the residue classes of permissible parts enumerated by both $LG_1^{**}(n)$ and $LG_2^{**}(n)$ are *asymmetric* modulo 8. This means that the corresponding infinite product generating functions are *not* modular forms. However, taken *together* a hidden symmetry is revealed, namely, $LG_1(n)$ counts partitions into parts congruent to 1, 5, or 6 modulo 8, while $LG_2(n)$ counts partitions into parts congruent to $8-1$, $8-5$, and $8-6$ modulo 8.

The analytic counterparts to Theorems 2.44 and 2.45 both arise as special cases of the Lebesgue identity (Theorem 1.74).

Theorem 2.47 (Analytic counterpart to Theorem 2.44).

$$\sum_{n=0}^{\infty} \frac{q^{n^2+n}(-q^{-1};q^2)_n}{(q^2;q^2)_n} = \frac{1}{(q;q^4)_\infty (q^6;q^8)_\infty}.$$

Proof. In Theorem 1.74, replace q by q^2 and set $a = -q^{-1}$. □

Theorem 2.48 (Analytic counterpart to Theorem 2.45).

$$\sum_{n=0}^{\infty} \frac{q^{n^2+n}(-q;q^2)_n}{(q^2;q^2)_n} = \frac{1}{(q^2;q^8)_\infty (q^3;q^4)_\infty}.$$

Proof. In Theorem 1.74, replace q by q^2 and set $a = -q$. □

The little Göllnitz identities can be proved from Theorems 2.44 and 2.45 in the same way that Theorem 2.42 was proved from Eq. (2.16).

2.4.6 BIG Göllnitz Partition Identity

Göllnitz [Göl67, Theorem 4.1] also discovered the following partition identity.

Theorem 2.49 (Big Göllnitz Identity). *Let $BG_1(n)$ denote the number of partitions of n into parts congruent to 2, 5, or 11 modulo 12. Let $BG_2(n)$ denote the number of partitions of n into distinct parts congruent to 2, 4, or 5 modulo 6. Let $BG_3(n)$ denote the number of partitions $\lambda = (\lambda_1, \lambda_2, \ldots, \lambda_{l(\lambda)})$ wherein*

- $\lambda_i \neq 1$ *and* $\lambda_i \neq 3$ *for any i,*

- $\lambda_i - \lambda_{i+1} \geqq 6$,

- $\lambda_i - \lambda_{i+1} > 6$ *if $i \equiv 0, 1, 3 \pmod 6$.*

Then $BG_1(n) = BG_2(n) = BG_3(n)$ for all n.

We notice immediately that the partitions enumerated by $BG_3(n)$ involve difference conditions more complicated than but nonetheless reminiscent of, partitions enumerated by the $s(n)$ in Schur's theorem, Theorem 2.36, in the sense that there are certain difference conditions that must always be obeyed, and stricter conditions that must be obeyed when the parts involved fall into certain residue classes. We also note that like the little Göllnitz identities, the partitions enumerated by $B_1(n)$ and $B_2(n)$ involve asymmetric residue classes. Our instincts tell us that Big Göllnitz must possess a companion identity. Indeed such a companion exists, but it was not discovered until half a century after Göllnitz's work, due in no small part to the unexpected initial condition involved. This companion identity is due to Alladi and Andrews [AA15].

Theorem 2.50 (Alladi–Andrews companion to Big Göllnitz). *Let $AA_1(n)$ denote the number of partitions of n into parts congruent to 1, 7, or 10 modulo 12. Let $AA_2(n)$ denote the number of partitions of n into distinct parts congruent to 1, 2, or 4 modulo 6. Let $AA_3(n)$ denote the number of partitions $\lambda = (\lambda_1, \lambda_2, \ldots, \lambda_{\ell(\lambda)})$ wherein*

* $\lambda_i - \lambda_{i+1} \geq 6$,

* $\lambda_i - \lambda_{i+1} > 6$ *if $i \equiv 0, 3, 5 \pmod 6$,*

with the exception that 5 and 1 may simultaneously appear as parts of λ. Then $AA_1(n) = AA_2(n) = AA_3(n)$ for all n.

Actually, we shall prove a refinement of Theorem 2.49, namely the following.

Theorem 2.51 (Refinement of Big Göllnitz identity). *Let $B(n, m)$ denote the number of partitions of n into m distinct parts $\equiv 2, 4, 5 \pmod 6$. Let $G(n, m)$ denote the number of partitions λ of n in which $\lambda_i - \lambda_{i+1} \geq 6$, $\lambda_i - \lambda_{i+1} > 6$ if $\lambda_1 \equiv 0, 1, 3 \pmod 6$, $m_1(\lambda) = m_3(\lambda) = 0$, and m denotes the number of parts $\equiv 2, 4, 6 \pmod 6$ plus twice the number of parts $\equiv 0, 1, 3 \pmod 6$.*

We shall follow closely a proof due to Andrews [And89, p. 105 ff.] (cf. [And86, Theorem 10.1, pp. 101–104]) which is conceptually simple but computationally cumbersome.

Proof. Let $G(n, m; N)$ denote the number of those partitions λ of the type enumerated by $G(n, m)$ in which $\lambda_1 \leq N$.
    Let
$$\mathscr{G}_N(t) := \mathscr{G}_N(t; q) := \sum_{n \geq 0} \sum_{m \geq 0} G(n, m; N) t^m q^n.$$

Thus the polynomials $\mathscr{G}_n(t)$ are completely determined by the recurrences

$$\mathscr{G}_{6n-a}(t) = \mathscr{G}_{6n-a-1}(t) + tq^{6n-a}\mathscr{G}_{6n-6-a}(t), n \geq 2, a \in \{2, 4, 5\}, \quad (2.82)$$
$$\mathscr{G}_{6n-a}(t) = \mathscr{G}_{6n-a-1}(t) + t^2 q^{6n-a}\mathscr{G}_{6n-7-a}(t), n \geq 2, a \in \{0, 1, 3\}, \quad (2.83)$$

and the initial conditions $\mathscr{G}_0(t) = \mathscr{G}_1(t) = 1$, $\mathscr{G}_2(t) = \mathscr{G}_3(t) = 1 + tq^2$, $\mathscr{G}_4(t) = 1 + tq^2 + tq^4$, and $\mathscr{G}_5(t) = 1 + tq^2 + tq^4 + tq^5$.

The key to the proof is showing that

$$\mathscr{G}_{6n+3}(t) - tq^{6n+2}\mathscr{G}_{6n-3}(t) = (1 + tq^2)(1 + tq^4)(1 + tq^5)\mathscr{G}_{6n-7}(tq^6), \quad (2.84)$$

where $\mathscr{G}_{-1}(t) = 1$.

For if (2.84) were true, then it would quickly follow that

$$\mathscr{G}_\infty(t) := \lim_{N \to \infty} \mathscr{G}_N(t) = \sum_{n,m \geq 0} G(n,m)t^m q^n, \quad (2.85)$$

whence by (2.84),

$$\mathscr{G}_\infty(t) = (1 + tq^2)(1 + tq^4)(1 + tq^5)\mathscr{G}_\infty(tq^6), \quad (2.86)$$

which, upon iteration, yields

$$\mathscr{G}_\infty(t) = (-tq^2; q^6)_\infty(-tq^4; q^6)_\infty(-tq^5; q^6)_\infty = \sum_{n,m \geq 0} B(n,m)t^m q^n. \quad (2.87)$$

Compare coëfficients of $t^m q^n$ in (2.87) with those of (2.85) to conclude that $B(n,m) = G(n,m)$.

Thus once we show that (2.84) holds, we will be finished.

Notice that the subsequence $\mathscr{G}_{6n-1}(t)$ of $\mathscr{G}_n(t)$ satisfies the following fourth order recurrence:

$$F(n, t, \mathscr{G}_{6n-1}(t), \mathscr{G}_{6n-7}(t), \mathscr{G}_{6n-13}(t), \mathscr{G}_{6n-19}(t), \mathscr{G}_{6n-25}(t)) = 0, \quad (2.88)$$

for $n \geq 3$, where

$$\begin{aligned} F(n, t, X, Y, Z, W, V) := &\ X - \left(1 + tq^{6n}(q^{-1} + q^{-2} + q^{-4})\right)Y \\ &- t^2 q^{6n}(q^{-3} + q^{-5} + q^{-6})(1 - q^{6n-6})Z \\ &- t^3 q^{12n-13}\left(t(q^{-1} + q^{-2} + q^{-4}) + q^{6n-12}\right)W - t^6 q^{18n-32}V, \quad (2.89) \end{aligned}$$

and we let $\mathscr{G}_{-1}(t) = -1$ and $\mathscr{G}_{-7}(t) = 0$. Similarly, one may show that

$$\mathscr{G}_{6n+3}(t) - tq^{6n+2}\mathscr{G}_{6n-3}(t) = H(n,t) \quad (2.90)$$

where

$$\begin{aligned} H(n,t) := &\ \mathscr{G}_{6n+5}(t) - tq^{6n+2}(1 + q^2 + q^3)\mathscr{G}_{6n-1}(t) \\ &+ t^2 q^{12n}(1 + q + q^3)\mathscr{G}_{6n-7}(t) - t^3 q^{18n-7}\mathscr{G}_{6n-13}(t). \quad (2.91) \end{aligned}$$

Using (2.88), the right-hand side of (2.84) satisfies

$$F\left(n-1, tq^6, \mathscr{G}_{6n-7}(tq^6), \mathscr{G}_{6n-13}(tq^6), \mathscr{G}_{6n-19}(tq^6), \mathscr{G}_{6n-25}(tq^6), \mathscr{G}_{6n-31}(tq^6)\right)$$
$$= 0 \quad (2.92)$$

for $n \geq 4$. For convenience, we let $J(n)$ denote the expression on the right-hand side of (2.88). We wish to show that for sufficiently large n,

$$F(n-1, tq^6, H(n,t), H(n-1,t), H(n-2,t), H(n-3,t), H(n-4,t)) = 0. \quad (2.93)$$

With the aid of our favorite computer algebra system, we write the left-hand side of (2.93) in terms of $J(n)$ by observing the coëfficients of $\mathscr{G}_{6n+5}(t)$, then $\mathscr{G}_{6n-1}(t)$, then $\mathscr{G}_{6n-7}(t)$, etc. Therefore

$$F(n-1, tq^6, H(n,t), H(n-1,t), H(n-2,t), H(n-3,t), H(n-4,t))$$
$$= J(n+1) - tq^{6n-6}(q^2 + q^4 + q^5)J(n) + t^2 q^{12n-12}(1 + q + q^3)J(n-1)$$
$$- t^3 q^{18n-25} J(n-2). \quad (2.94)$$

We have $J(n) = 0$ for $n \geq 3$, and thus (2.93) is true for $n \geq 5$.

And so, by (2.92) and (2.93), both sides of (2.84) satisfy the same fourth order recurrence for $n \geq 5$. It is easily checked via a computer algebra system that (2.84) is satisfied for $1 \leq n \leq 4$. Accordingly, (2.84) is established, and thus also Theorem (2.51) and Theorem (2.49). $\qquad \square$

2.4.7 Exercises

1. Prove Theorem 2.34.

2. Prove Theorem 2.43.

2.5 Close Cousins of Rogers–Ramanujan Type Identities

2.5.1 False Theta Functions

L. J. Rogers coined the term *false theta function* and provided a number of examples in his paper [Rog17]. False theta functions are like Rogers–Ramanujan type series, but with some of the plus and minus signs reversed. Specifically, recalling that Ramanujan's theta series

$$f(a,b) := \sum_{n=-\infty}^{\infty} a^{n(n+1)/2} b^{n(n-1)/2} =$$

$$\sum_{n=0}^{\infty} a^{n(n+1)/2} b^{n(n-1)/2} + \sum_{n=1}^{\infty} a^{n(n-1)/2} b^{n(n+1)/2},$$

we may define the corresponding *false theta series* as

$$\Psi(a,b) := \sum_{n=0}^{\infty} a^{n(n+1)/2} b^{n(n-1)/2} - \sum_{n=1}^{\infty} a^{n(n-1)/2} b^{n(n+1)/2}.$$

Notice that although it is always the case that $f(a,b) = f(b,a)$, usually $\Psi(a,b) \neq \Psi(b,a)$. L. J. Rogers [Rog17] studied q-series expansions for many instances of $f(\pm q^r, \pm q^s)$ and $\Psi(\pm q^r, \pm q^s)$, which he called *theta (resp. false theta) series of order* $(r+s)/2$. Ramanujan was aware of Rogers' work on false theta functions and recorded a number of identities involving false theta functions in the lost notebook. A false theta series identity for the series $\Psi(\pm q^r, \pm q^s)$ arises from the same Bailey pair as the Rogers–Ramanujan type identity with product $f(\pm q^r, \pm q^s)/\varphi(-q)$.

Identities involving false theta functions may be obtained by the following limiting case of Bailey's lemma:

Corollary 2.52. *If $(\alpha_n(q,q), \beta_n(q,q))$ forms a Bailey pair with respect to q, then*

$$\frac{1}{1-q} \sum_{n=0}^{\infty} (-1)^n q^{n(n+1)/2} (q;q)_n \beta_n(q,q) = \sum_{r=0}^{\infty} q^{r(r+1)/2} \alpha_r(q,q). \qquad (2.95)$$

Proof. To obtain (2.95), set $x = \rho_2 = q$ in (2.38). $\qquad \square$

Consider the following example.

Lemma 2.53. *If*

$$\alpha_n = q^{n^2+n/2} \frac{1+q^{n+1/2}}{1+\sqrt{q}}$$

and

$$\beta_n = \frac{1}{(q^{3/2};q)_n (q;q)_n}$$

then (α_n, β_n) forms a Bailey pair with respect to q.

Proof. By inserting the above pair into (2.29), the identity we need to establish is

$$\frac{1}{(q;q)_n (q^{3/2};q)_n} = \sum_{r=0}^{n} \frac{q^{r^2+r/2}(1+q^{r+1/2})}{(q;q)_{n-r}(q^2;q)_{n+r}(1+q^{1/2})},$$

which can be verified by the q-Zeilberger algorithm. $\qquad \square$

Theorem 2.54.

$$\sum_{n=0}^{\infty} \frac{q^{n(n+1)}(-q^2;q^2)_n}{(q;q)_{2n+1}} = \frac{f(q,q^5)}{\varphi(-q^2)} \qquad (2.96)$$

and

$$\sum_{n=0}^{\infty} \frac{(-1)^n q^{n(n+1)}(q^2;q^2)_n}{(q;q)_{2n+1}} = \Psi(-q^5,-q). \qquad (2.97)$$

Proof. To obtain (2.96) and (2.97), insert the Bailey pair in Lemma 2.53, with q replaced by q^2, into (2.42) and (2.95), respectively, again with q replaced by q^2. □

Equation (2.97) is due to Rogers [Rog17, p. 333, Eq. (4)]. Equation (2.96) was recorded by Ramanujan in the lost notebook [AB05, p. 254, Eq. (11.3.5)] and proved by Slater [Sla52, p. 154, Eq. (28)]. Unlike their close cousins, the mock theta functions which have received a steady stream of attention since their discovery, after their initial discovery by Rogers in 1917 and some early contributions recorded by Ramanujan in his lost notebook, the false theta functions have only received sporadic attention until quite recently. In the 1940s, Dyson contributed two false theta function identities to Bailey's paper [Bai47, p. 434, Eqs. (E1) and (E2)]. In his 1979 monograph *Partitions: Yesterday and Today*, Andrews provided an extended discussion of the false theta functions, including partition theoretic interpretations [And79, pp. 42–55]. McLaughlin, Sills, and Zimmer gave some new false theta function identities in the spirit of Rogers and Ramanujan in [MS08, MSZ09]. In [War03], Warnaar shows how the false theta functions fit into the more general theory of partial theta functions. After the introduction of "quantum modular forms" by Don Zagier in his paper [Zag10], there has been significant progress toward understanding the connections between false theta functions and classical automorphic forms; see e.g. the paper by Folsom, Ono, and Rhoades [FOR13].

2.5.2 Mock Theta Functions

In his famous and oft-quoted last letter to Hardy [BR95, pp. 220–223], [Ram27, pp. 354–355], [Wat36, p. 57–61] dated 12 January 1920[3], Ramanujan introduces what he calls the *mock theta functions*. Ramanujan wrote:

> I discovered very interesting functions recently which I call "Mock" ϑ-functions. Unlike the "False" ϑ-functions (studied partially by Prof. Rogers in his interesting paper) they enter into mathematics as beautifully as the ordinary ϑ-functions. I am sending you with this letter some examples...

Ramanujan does not give an explicit definition of mock theta function, but rather describes some of the properties that such functions should have, and offers a number of examples, without proving that his examples actually satisfy the required properties. (Griffin, Ono, and Rolen proved that Ramanujan's mock theta functions *do* in fact satisfy these properties in [GOR13].) Until the unearthing of Ramanujan's lost notebook in 1976 by George Andrews, the only source of Ramanujan's thoughts on mock theta functions was this single letter to Hardy. Mock theta functions have captured the imagination

---

[3]A photographic copy of the final five pages of this letter written in Ramanujan's own hand is reproduced in [Ram88, pp. 127–131].

of many first-rate mathematicians over the decades, so they have a vast and extensive literature. The earliest work (after Ramanujan's death) on the mock theta functions is due to Watson [Wat36, Wat37]. Two of Rademacher's students, Leila Dragonette and George Andrews, worked on the mock theta functions for their PhD dissertations. Dean Hickerson proved the so-called "mock theta conjectures" in [Hic88]. The major breakthrough that frames all current work on the mock theta functions was due to Sander Zwegers, who, in his 2002 PhD thesis [Zwe02] showed that all of Ramanujan's examples of mock theta functions fit naturally into a theory of harmonic Maass forms. Katherin Bringmann, Amanda Folsom, Ken Ono, Rob Rhoades, Don Zagier, and others have made contributions to the understanding of mock theta functions in recent years. (See [Duk14] for plenty of references, which in turn point to numerous other references.)   We shall make no attempt to seriously discuss the mock theta functions here; the mock theta functions deserve an entire book of their own! (See *Harmonic Maass Forms and Mock Modular Forms: Theory and Applications*, by Bringmann, Folsom, Ono, and Rolen.)   For a short, accessible introduction to the mock theta functions, see the *Notices* article by William Duke [Duk14]. For a discussion of Ramanujan's (implied) definition of mock theta function, and the modern (i.e. post-Zwegers) definition, and how they are not entirely compatible, see the *PNAS* article by Rob Rhodes [Rho13].

Here we shall merely display some of Ramanujan's examples, so that the reader may observe the similarity to Rogers–Ramanujan type identities.

We begin with eight of the ten "fifth order" mock theta functions given by Ramanujan in his letter to Hardy. Ramanujan does not define what he means by "order," but presumably he means that these functions are in some sense counterparts to Rogers–Ramanujan type functions in which the infinite product representation involves $f(\pm q^a, \pm q^b)$, where $a + b = 5$. The notion of order of a mock theta function was put on firm ground and explicitly defined as a result of the work of Zwegers. Clearly, Eqs. (2.98) and (2.102) are respectively the mock theta counterparts to the first and second Rogers–Ramanujan identities (2.3) and (2.4). The function names and summation representations are Ramanujan's own (subscripts were added by Watson to distinguish the two sets of five each); the alternate representations as Hecke-type double series were given by Andrews [And86, p. 125, Theorem 9].

$$f_0(q) := \sum_{n=0}^{\infty} \frac{q^{n^2}}{(-q;q)_n} = \frac{1}{(q;q)_\infty} \sum_{n\geq 0} \sum_{|j|\leq n} (-1)^j q^{n(5n+1)/2-j^2}(1 - q^{4n+2}),$$

$$(2.98)$$

$$F_0(q) := \sum_{n=0}^{\infty} \frac{q^{2n^2}}{(q;q^2)_n} = \frac{1}{(q^2;q^2)_\infty} \sum_{n=0}^{\infty} \sum_{j=0}^{2n}(-1)^n q^{5n^2+2n-j(j+1)/2}(1 + q^{6n+3}),$$

$$(2.99)$$

$$1 + 2\psi_0(q) := \sum_{n=0}^{\infty} q^{n(n+1)/2}(-1;q)_n$$

$$= \frac{1}{\varphi(-q)}$$

$$\times \left(1 + 2\sum_{n=1}^{\infty}(-1)^n q^{2n^2+n} - 2\sum_{n=1}^{\infty}\sum_{|j|<n}(-1)^j q^{n(5n-1)/2-j(3j+1)/2}(1-q^n) \right),$$
$$\tag{2.100}$$

$$\phi_0(q) := \sum_{n=0}^{\infty}(-q;q^2)_n q^{n^2}$$

$$= \frac{1}{\psi(-q)}\sum_{n=0}^{\infty}\sum_{|j|\leq n}(-1)^j q^{n(5n+2)-j(3j+1)}(1-q^{6n+3}), \quad (2.101)$$

$$f_1(q) := \sum_{n=0}^{\infty}\frac{q^{n(n+1)}}{(-q;q)_n} = \frac{1}{(q;q)_{\infty}}\sum_{n=0}^{\infty}\sum_{|j|\leq n}(-1)^j q^{n(5n+3)-j^2}(1+q^{2n+1}),$$
$$\tag{2.102}$$

$$F_1(q) := \sum_{n=0}^{\infty}\frac{q^{2n(n+1)}}{(q;q^2)_{n+1}} = \frac{1}{(q^2;q^2)_{\infty}}\sum_{n=0}^{\infty}\sum_{j=0}^{2n}(-1)^n q^{5n^2+4n-j(j+1)/2}(1+q^{2n+1}),$$
$$\tag{2.103}$$

$$\psi_1(q) = \sum_{n=0}^{\infty}(-q;q)_n q^{n(n+1)/2}$$

$$= \frac{1}{\varphi(-q)}\sum_{n=0}^{\infty}\sum_{|j|\leq n}(-1)^j q^{n(5n+3)/2-j(3j+1)/2}(1-q^{2n+1}), \quad (2.104)$$

$$\phi_1(q) := \sum_{n=0}^{\infty}(-q;q^2)_n q^{(n+1)^2}$$

$$= \frac{1}{\psi(-q)}\sum_{n=0}^{\infty}\sum_{|j|\leq n}(-1)^j q^{n(5n+4)-j(3j+1)}(1-q^{2n+1}). \quad (2.105)$$

Ramanujan also gave two other fifth order mock theta functions, for which Andrews could not find representations as Hecke-type double series. In [Zwe09, p. 208, Theorem 1], Zwegers gives the analogous representations as triple series:

$$\chi_0(q) := \sum_{n=0}^{\infty}\frac{q^n}{(q^{n+1};q)_n} =$$

$$2 - \frac{1}{(q^2; q^2)_\infty} \left(\sum_{k,l,m \geq 0} - \sum_{k,l,m < 0} \right) (-1)^{k+l+m} q^{Q(k,l,m) + \frac{1}{2}(k+l+m)}, \quad (2.106)$$

$$\chi_1(q) := \sum_{n=0}^{\infty} \frac{q^n}{(q^{n+1}; q)_{n+1}} =$$

$$\frac{1}{(q^2; q^2)_\infty} \left(\sum_{k,l,m \geq 0} - \sum_{k,l,m < 0} \right) (-1)^{k+l+m} q^{Q(k,l,m) + \frac{3}{2}(k+l+m)}, \quad (2.107)$$

where $Q(k, l, m) = \frac{1}{2}k^2 + \frac{1}{2}l^2 + \frac{1}{2}m^2 + 2kl + 2km + 2lm$.

Let us take a closer look at the identity (2.98) for $f_0(q)$. Notice that the single-sum representation (the representation given by Ramanujan in what he called the "Eulerian" form) is just like the sum side of the first Rogers–Ramanujan identity but with a minus sign in front of q in the denominator rising q-factorial. Recalling that the first Rogers–Ramanujan identity may be derived by inserting a certain Bailey pair, the one in which $\beta_n = 1/(q; q)_n$, into (2.39).

It would be therefore reasonable to suspect that (2.98) may result from inserting the Bailey pair with $\beta_n = 1/(-q; q)_n$ into (2.39).

To get the α_n associated with $\beta_n = 1/(-q; q)_n$, we can use Eq. (2.31) with $x = 1$ and let our favorite computer algebra system show that

$$\alpha_0 = 1$$
$$\alpha_1 = q(q - 3)$$
$$\alpha_2 = q^3(q^4 - 2q^3 - q^2 + 2q + 2)$$
$$\alpha_3 = q^6(q^9 - 2q^8 - q^6 + 4q^5 - 2q^2 - 2)$$
$$\alpha_4 = q^{10}(q^{16} - 2q^{15} + q^{12} + 2q^{11} - 2q^8 - 2q^7 + 2q^3 + 2)$$
$$\alpha_5 = q^{15} \left(q^{25} - 2q^{24} + 2q^{21} - q^{20} + 2q^{19} - 4q^{16} + 2q^{11} + 2q^9 - 2q^4 - 2 \right)$$
$$\alpha_6 = q^{21} \left(q^{36} - 2q^{35} + 2q^{32} - q^{30} + 2q^{29} - 2q^{27} - 2q^{26} + 2q^{21} + 2q^{20} - 2q^{14} \right.$$
$$\left. -2q^{11} + 2q^5 + 2 \right).$$

A careful look at the above list (as well as α_7, α_8, and α_9 which are not reproduced here) was enough to lead George Andrews to conjecture:

$$\alpha_n = q^{n(3n+1)/2} \sum_{j=-n}^{n} (-1)^j q^{-j^2} - q^{n(3n-1)/2} \sum_{j=-n+1}^{n-1} (-1)^j q^{-j^2}. \quad (2.108)$$

Upon establishing that the α_n in (2.108) does indeed form a Bailey pair with $\beta_n = 1/(-q; q)_n$, insertion of this pair into (2.39) yields (2.98). The other mock theta function identities can be established similarly.

2.6 Enumeration

Thus far, we have seen that the first Rogers–Ramanujan identity implies that the number of partitions of n into distinct, nonconsecutive parts equals the number of partitions of n into parts $\equiv \pm 1 \pmod 5$. It is only natural to further ask, "How many such partitions are there?" Is there a direct formula for this number of partitions?

Before we can answer this question, we should ask the more fundamental question, "how many *unrestricted* partitions of n are there?" Euler answered this question via a recurrence, Theorem 1.22. To get an exact evaluation of $p(n)$ using Euler's recurrence, one has to know the values of $p(n-1), p(n-2), p(n-5), p(n-7), \ldots, p(n-j(3j-1)/2), p(n-j(3j+1)/2), \ldots$, stopping only when the argument of $p(n)$ becomes negative. The total number of nonzero terms will thus be about $\frac{2}{3}\sqrt{6n}$. (**Exercise:** Prove the statement in the preceding sentence.)

But what about a direct, non-recursive formula? The first to study this question were Hardy and Ramanujan.

Remark 2.55. Here we need to temporarily depart from our philosophy that q is a formal variable for the remainder of this section. The analytic techniques employed to derive formulæ such as those about to be presented require us to consider the associated generating functions not as formal power series or formal infinite products, but rather as functions of the complex variable q, absolutely convergent when $|q| < 1$. Furthermore, it will be convenient to adopt the standard asymptotic notations:

$$f(n) \sim g(n)$$

as $n \to \infty$ to mean

$$\lim_{n\to\infty} \frac{f(n)}{g(n)} = 1,$$

and

$$f(n) = O(g(n))$$

as $n \to \infty$ to mean that there exists a positive real constant M and a positive integer N such that

$$\left| \frac{f(n)}{g(n)} \right| \leqq M$$

whenever $n \geqq N$.

2.6.1 Asymptotic formulæ

Before considering exact formulæ we will consider asymptotic formulæ. Perhaps the most famous asymptotic formula in the theory of numbers is the *prime number theorem*, conjectured by Gauss as a teenager, and finally proved a bit more than a century later independently by Jacques Hadamard [Had96] and Charles Jean de la Vallé-Poussin [dlVP96] in 1896. The prime number theorem states that if $\pi(n)$ denotes the number of primes less than or equal to n, then

$$\pi(n) \sim \frac{n}{\log n} \text{ as } n \to \infty.$$

2.6.1.1   Partition Function

Given the excitement in the world of analytic number theory generated by the proof of the prime number theorem at the close of the nineteenth century, it is not surprising that, a couple of decades later, Hardy and Ramanujan [HR18, p. 79, Eq. (1.41)] (and later, independently, J. V. Uspensky [Usp20]) examined an analogous question in additive number theory and showed

$$p(n) \sim \frac{1}{4n\sqrt{3}} e^{\pi\sqrt{2n/3}} \text{ as } n \to \infty. \tag{2.109}$$

A somewhat more precise asymptotic formula is given by Marvin I. Knopp [Kno93, p. 90, Theorem 2]:

$$p(n) = \frac{1}{4\sqrt{3}\left(n - \frac{1}{24}\right)} \exp\left(\pi\sqrt{\frac{2}{3}\left(n - \frac{1}{24}\right)}\right)\left(1 - \frac{1}{\pi\sqrt{\frac{2}{3}\left(n - \frac{1}{24}\right)}}\right)$$
$$+ O\left(\exp\left(\pi\sqrt{\frac{1}{6}\left(n - \frac{1}{24}\right)}\right)\right).$$

2.6.1.2   Meinardus' Theorem and Todt's Extension

It turns out that (2.109) is included as a special case of a much more general theorem, which we will now examine in some detail.

Let

$$F(q) := F(q; \{a_n\}_{n=1}^\infty) = \prod_{n=1}^\infty \frac{1}{(1-q^n)^{a_n}} =: \sum_{n=0}^\infty r(n)q^n,$$

where here and throughout, $|q| < 1$, and (for the moment) $\{a_n\}_{n=0}^\infty$ is a sequence of nonnegative integers.

Next, use the $\{a_n\}$ to build an associated Dirichlet series

$$D(s) := \sum_{n=1}^\infty \frac{a_n}{n^s},$$

convergent for $\Re s > \alpha$, and possessing an analytic continuation to $\Re s \geq -C_0$,

$(0 < C_0 < 1)$ and that in this region $D(s)$ is analytic except for a pole of order 1 at $s = \alpha$ with residue A. Also,

$$D(s) = O(|\Im s|^{C_1})$$

uniformly in $\Re s \geq -C_0$ as $|\Im s| \to \infty$.

Günter Meinardus [Mei54] proved the following result:

Theorem 2.56. *Given the definitions above, as $n \to \infty$,*

$$r(n) \sim Cn^\kappa \exp\left\{ n^{\alpha/(\alpha+1)} \left(1 + \frac{1}{\alpha}\right) [A\Gamma(\alpha+1)\zeta(\alpha+1)]^{1/(\alpha+1)} \right\},$$

where $\zeta(s)$ is the Riemann zeta function, $\Gamma(s)$ is Euler's gamma function, and

$$C = \frac{e^{D'(0)}}{\sqrt{2\pi(1+\alpha)}} (A\Gamma(\alpha+1)\zeta(\alpha+1))^{(1-2D(0))/(2+2\alpha)},$$

$$\kappa = \frac{2D(0) - 2 - \alpha}{2(1+\alpha)}.$$

See [And76, Chapter 6] for an exposition of Theorem 2.56 in English. Heiko Todt [Tod11] extended Theorem 2.56 to allow for the $\{a_n\}$ to be a sequence of any real numbers, provided $r(n)$ is a nondecreasing function of n.

2.6.1.3 The Case of Rogers–Ramanujan–Slater Products

Two years before the appearance of Meinardus' paper [Mei54], Slater published her paper [Sla52] containing a list of 130 identities of Rogers–Ramanujan type. Those infinite products $F(q)$ that arise in Rogers–Ramanujan–Slater type identities have the property that the a_n are a periodic function of n. Discrete periodic functions can be written in terms of powers of complex roots of unity. However, let us adopt a more straightforward notation here. Define the *circulator function modulo m* as

$$\mathrm{crm}_n(a_1, a_2, \ldots, a_m) := a_j$$

when $n \equiv j \pmod{m}$. Furthermore, the average of the a_j in a complete period occurs frequently enough that we will define the symbols

$$\mathbf{a} := (a_1, a_2, \ldots, a_m)$$

and

$$\bar{\mathbf{a}} := \frac{1}{m} \sum_{j=1}^{m} a_j.$$

Remark 2.57. Circulator functions (albeit with a different notation than is employed here) appear to have been introduced by J. F. W. Herschel in [Her18].

Thus, for example, the a_n associated with the first Rogers–Ramanujan identity, in which the infinite product is

$$\prod_{k=1}^{\infty} \frac{1}{(1 - q^{5k-4})(1 - q^{5k-1})},$$

is given by $a_n = \mathrm{cr}5_n(1, 0, 0, 1, 0)$.

Further, we note that in the case of most Rogers–Ramanujan–Slater type identities, the a_n are *periodically symmetric*, meaning that $a_j = a_{m-j}$ for $j = 1, 2, \ldots, m - 1$.

Next, we consider the associated Dirchlet series $D(s)$. Observe that

$$D(s) = \frac{1}{m^s} \sum_{j=1}^{m} a_j \zeta\left(s, \frac{j}{m}\right),$$

where $\zeta(s, t)$ is the *Hurwitz zeta function*, i.e., the analytic continuation of the function initially defined by the infinite series

$$\sum_{n=0}^{\infty} (t + n)^{-s},$$

which is defined for all complex $s \neq 1$. At $s = 1$, $\zeta(s, t)$ has a simple pole with residue 1. Given that $D(s)$ has such a nice form when a_n is periodically symmetric, we may proceed to simplify the statement of the theorem for these important special cases. To begin, $\alpha = 1$, and this implies the following: the residue A of $D(s)$ at $s = 1$ is given by

$$A = \frac{1}{m} \sum_{j=1}^{m} a_j = \bar{a},$$

and we recall well-known facts that $\Gamma(2) = 1$ and $\zeta(2) = \pi^2/6$.

Also, by Leibniz' rule from elementary calculus,

$$D'(s) = \frac{1}{m^s} \sum_{j=1}^{m} a_j \zeta'\left(s, \frac{j}{m}\right) - \frac{\log m}{m^s} \sum_{j=1}^{m} a_j \zeta\left(s, \frac{j}{m}\right)$$

$$= \frac{1}{m^s} \sum_{j=1}^{m} a_j \zeta'\left(s, \frac{j}{m}\right) - \log m \, D(s),$$

where all derivatives are with respect to s.

But we only need the values of $D(s)$ and $D'(s)$ at $s = 0$. Recalling [Apo76, p. 268] that $\zeta(0, t) = \frac{1}{2} - t$ provided $\Re t > 0$, we find that

$$D(0) = \sum_{j=1}^{m} a_j \left(\frac{1}{2} - \frac{j}{m}\right)$$

$$= \frac{1}{2}(a_1 + a_2 + \cdots + a_m) - \frac{1}{m}(a_1 + 2a_2 + 3a_3 + \cdots + ma_m)$$

$$= -\frac{a_m}{2} + \frac{1}{2m} \sum_{j=1}^{m-1} (m - 2j)a_j,$$

where we note that $\sum_{j=1}^{m-1}(m-2j)a_j = 0$ whenever the $\{a_n\}$ are periodically symmetric. So in the case of periodic symmetry of the $\{a_n\}$, we have

$$D(0) = -\frac{a_m}{2}.$$

We shall assume periodic symmetry of the $\{a_n\}$ from here on.

Next, we use the well-known fact that $\zeta'(0,t) = \log\Gamma(t) - \frac{1}{2}\log 2\pi$, and the above to deduce

$$D'(0) = \frac{a_m}{2}\log m + \log\left(\prod_{j=1}^{m}\left(\frac{\Gamma(j/m)}{\sqrt{2\pi}}\right)^{a_j}\right),$$

and thus, recalling Euler's reflection formula,

$$\Gamma(s)\Gamma(1-s) = \pi\csc(\pi s),$$

we have

$$e^{D'(0)} = \frac{m^{a_m/2}}{(2\pi)^{\frac{1}{2}m\bar{a}}} \prod_{j=1}^{m} (\Gamma(j/m))^{a_j}$$

$$= \frac{m^{a_m/2}}{(2\pi)^{\frac{1}{2}m\bar{a}}} \prod_{j=1}^{\lfloor m/2\rfloor} \pi^{a_j}\csc^{a_j}\left(\frac{j\pi}{m}\right)$$

$$= \frac{m^{a_m/2}}{2^{\frac{1}{2}m\bar{a}}} \prod_{j=1}^{\lfloor (m-1)/2\rfloor} \csc^{a_j}\left(\frac{j\pi}{m}\right).$$

Thus, we now conclude that

$$C = \frac{m^{a_m/2}\bar{a}^{(1+a_m)/4}}{2^{\frac{1}{2}m\bar{a}}\pi^{a_m}2^{(5+a_m)/4}\cdot 3^{(1+a_m)/4}} \prod_{j=1}^{\lfloor\frac{m-1}{2}\rfloor} \csc^{a_j}\left(\frac{j\pi}{m}\right),$$

and

$$\kappa = -\frac{a_m + 3}{4}.$$

Therefore, we have the following:

Corollary 2.58. *If $a_n = crm_n(a_1, a_2, \ldots, a_m)$ and $a_j = a_{m-j}$ for $j = 1, 2, \ldots, m-1$, then*

$$r(n) \sim \frac{m^{a_m/2}}{2^{1+\frac{m\bar{a}}{2}}}\left(\frac{\bar{a}}{6}\right)^{\frac{1+a_m}{4}} \prod_{j=1}^{\lfloor\frac{m-1}{2}\rfloor} \csc^{a_j}\left(\frac{j\pi}{m}\right) n^{-(a_m+3)/4}\exp\left(\pi\sqrt{\frac{2n\bar{a}}{3}}\right)$$

as $n \to \infty$.

2.6.1.4   Examples of Applications of Meinardus' Theorem

Asymptotic formulæ for many combinatorial functions may be deduced from Meinardus' theorem and Todt's extension. Often these special cases reduce to much simpler forms. Some examples follow.

Example 2.59 (partitions into distinct parts). Since, by Theorem 1.8,

$$\sum_{n=0}^{\infty} pd(n)q^n = \prod_{n=1}^{\infty} \frac{1}{(1-q^n)^{a_n}}$$

with

$$a_n = \mathrm{cr}2_n(1,0) = \chi_1^{(2)}(n),$$

where $\chi_1^{(2)}(n)$ is the principal Dirichlet character modulo 2, we have

$$pd(n) \sim \frac{1}{4 \cdot 3^{1/4} n^{3/4}} e^{\pi \sqrt{n/3}}$$

as $n \to \infty$.

Example 2.60 (unrestricted overpartitions). We have seen that

$$\sum_{n \geq 0} \overline{p}(n)q^n = \frac{(-q;q)_\infty}{(q;q)_\infty} = \prod_{n=1}^{\infty} \frac{1}{(1-q^n)^{a_n}}$$

with $a_n = \mathrm{cr}2_n(2,1)$. Thus as $n \to \infty$,

$$\overline{p}(n) \sim \frac{1}{8n} e^{\pi \sqrt{n}}.$$

Example 2.61 (Partitions with no repeated odd parts). Since

$$\sum_{n \geq 0} pod(n)q^n = \frac{(-q;q^2)_\infty}{(q^2;q^2)_\infty} = \prod_{n=1}^{\infty} \frac{1}{(1-q^n)^{a_n}}$$

with $a_n = \mathrm{cr}4_n(1,0,1,1)$, and thus

$$pod(n) \sim \frac{1}{4n\sqrt{2}} e^{\pi \sqrt{n/2}}$$

as $n \to \infty$.

Example 2.62 (Rogers–Ramanujan partitions). For $j = 1, 2$,

$$\sum_{n \geq 0} r_j(n)q^n = \frac{1}{(q^j;q^5)_\infty (q^{5-j};q^5)_\infty}, \tag{2.110}$$

and so as $n \to \infty$,

$$r_j(n) \sim \frac{\csc(j\pi/5)}{4 \cdot 15^{1/4} n^{3/4}} \exp\left(\frac{2\pi \sqrt{n}}{\sqrt{15}}\right).$$

Example 2.63 (Göllnitz–Gordon partitions). For $j = 1, 3$,

$$\sum_{n=0}^{\infty} G_j(n)q^n = \frac{1}{(q^j, q^4, q^{8-j}; q^8)_\infty},$$

and so as $n \to \infty$,

$$G_j(n) \sim \frac{\csc(j\pi/8)}{8\sqrt{2}\, n^{3/4}} \exp\left(\frac{\pi}{2}\sqrt{n}\right).$$

Example 2.64 (Schur's partitions).

$$\sum_{n\geq 0} s(n)q^n = \frac{1}{(q;q^6)_\infty (q^5;q^6)_\infty} = \prod_{n=1}^{\infty} \frac{1}{(1-q^n)^{a_n}},$$

$$a_n = \mathrm{cr}6_n(1,0,0,0,1,0) = \chi_1^{(6)}(n),$$

where $\chi_1^{(6)}(n)$ is the principal Dirichlet character modulo 6. As $n \to \infty$,

$$s(n) \sim \frac{1}{2 \cdot 2^{3/4}\sqrt{3}n^{3/4}} \exp\left(\frac{\pi\sqrt{2n}}{3}\right).$$

Example 2.65 (Coëfficients of the Rogers–Selberg series). Recall the three Rogers–Selberg identities (2.11), (2.12), (2.13) where the products are

$$\frac{f(-q^j, -q^{7-j})}{(q^2;q^2)_\infty} =: \sum_{n=0}^{\infty} rs_j(n)q^n,$$

for $j = 1, 2, 3$. Meinardus' theorem and even Todt's extension do not apply directly, as the series expansions involve an alternating sign. However, if we replace q by $-q$, and instead consider the expansions of

$$\frac{f(q, -q^6)}{(q^2;q^2)_\infty} = \sum_{n=0}^{\infty} |rs_1(n)|q^n,$$

$$\frac{f(-q^2, q^5)}{(q^2;q^2)_\infty} = \sum_{n=0}^{\infty} |rs_2(n)|q^n,$$

$$\frac{f(q^3, -q^4)}{(q^2;q^2)_\infty} = \sum_{n=0}^{\infty} |rs_3(n)|q^n,$$

the conditions of Todt's extension are now satisfied. Thus, we may conclude that as $n \to \infty$,

$$|rs_j(n)| \sim \frac{\sqrt[4]{\frac{11}{21}}}{64\, 2^{3/4}n^{3/4}} \exp\left(\pi\sqrt{\frac{11n}{42}}\right) \prod_{k=1}^{6} \csc\left(b_{j,k}\frac{\pi}{28}\right),$$

for $j = 1, 2, 3$; where

$$(b_{1,1}, b_{1,2}, b_{1,3}, b_{1,4}, b_{1,5}, b_{1,6}) = (1, 4, 7, 10, 12, 13),$$
$$(b_{2,1}, b_{2,2}, b_{2,3}, b_{2,4}, b_{2,5}, b_{2,6}) = (4, 5, 6, 7, 8, 9),$$
$$(b_{3,1}, b_{3,2}, b_{3,3}, b_{3,4}, b_{3,5}, b_{3,6}) = (2, 3, 7, 8, 11, 12).$$

Of course,

$$rs_j(n) = (-1)^n |rs_j(n)|.$$

2.6.2   Exact Formulæ for Partition Enumeration Functions

2.6.2.1 $p(n)$

Hardy and Ramanujan [HR18, p. 85, Eq. (1.75)] further gave an asymptotic formula for $p(n)$ with sufficient precision so that by rounding to the nearest integer, $p(n)$ may be found *exactly*:

$$p(n) = \left\lfloor \frac{1}{2\pi\sqrt{2}} \sum_{k=1}^{\lfloor \alpha\sqrt{n} \rfloor} \sqrt{k} A_k(n) \frac{d}{dn} \left(\frac{\exp\left(\sqrt{\frac{2}{3}\left(n - \frac{1}{24}\right)}/k \right)}{\sqrt{n - \frac{1}{24}}} \right) \right\rceil, \qquad (2.111)$$

where $\lfloor x \rceil$ is the nearest integer to x, α is a real constant, $A_k(n)$ is a Kloosterman-type sum given by

$$A_k(n) := \sum_{h \in \mathbb{Z}_k^\times} \omega(h, k) e^{2\pi i h/k},$$

and $\omega(h, k)$ is a 24kth root of unity[4] given by

$$\omega(h, k)$$
$$:= \begin{cases} \left(\frac{-k}{h}\right) \exp\left[-\left\{\frac{1}{4}(2 - hk - h) + \frac{1}{12}\left(k - \frac{1}{k}\right)(2h - h' + h^2 h')\right\}\pi i\right] & \text{if } 2 \nmid h \\ \left(\frac{-h}{k}\right) \exp\left[-\left\{\frac{1}{4}(k - 1) + \frac{1}{12}\left(k - \frac{1}{k}\right)(2h - h' + h^2 h')\right\}\pi i\right] & \text{if } 2 \nmid k \end{cases}$$

In the Fall Semester of 1936, while preparing to present the derivation of the Hardy–Ramanujan formula for $p(n)$ to his graduate number theory class, Hans Rademacher discovered that a slight change in Hardy and Ramanujan's analysis allowed him to produce a series that converges to $p(n)$:

$$p(n) = \frac{1}{\pi\sqrt{2}} \sum_{k \geq 1} \sqrt{k}\, A_k(n) \frac{d}{dn} \left(\frac{1}{\sqrt{n - \frac{1}{24}}} \sinh\left(\frac{\pi}{k}\sqrt{\frac{2}{3}\left(n - \frac{1}{24}\right)} \right) \right).$$
$$(2.112)$$

---

[4]Tom Apostol showed in his 1948 PhD thesis [Apo48] that $\omega(h, k)$ is in fact also a 12kth root of unity, but everyone seems to ignore this fact, and so $\omega(h, k)$ continues to be referred to as 24kth root of unity in the literature.

The method for deriving these asymptotic formulæ and convergent series for $p(n)$, which is applicable to many problems in analytic number theory, is known as the *circle method*. The circle method originated in the work of Hardy and Ramanujan, was developed further by Hardy and Littlewood, and used extensively by Rademacher, his students, and many others.

The derivation of (2.112) via the circle method will not be done here. The interested reader is referred to the expositions by Rademacher [Rad37, Rad43, Rad73], Andrews [And76, Chapter 5], and Apostol [Apo97, Chapter 5].

We note that (2.112) may also be written as

$$p(n) = 2\pi(24n-1)^{-3/4} \sum_{k\geq 1} \frac{A_k(n)}{k} I_{3/2}\left(\frac{\pi\sqrt{24n-1}}{6k}\right), \tag{2.113}$$

in which I_ν is a *modified Bessel function of the first kind of order* ν,

$$I_\nu(z) := \sum_{n\geq 0} \frac{1}{n!\,\Gamma(n+\nu+1)} \left(\frac{z}{2}\right)^{2n+\nu},$$

where z is a purely imaginary number, i.e. $\Re z = 0$. Bessel functions arise naturally in the study of differential equations; the definitive reference on Bessel functions was written by Watson [Wat44]. It turns out that Bessel functions of half-odd order can be written in terms of elementary functions, allowing the form given in (2.112).

The derivation of (2.112) via circle method relies heavily on the fact that the generating for $p(n)$ is (up to a trivial multiple) a modular form.

In late 2010, Jan Bruinier and Ken Ono [BO13], found a formula which expresses $p(n)$ as a finite sum of algebraic numbers. These algebraic numbers are the singular moduli for $F(\tau)$ defined by

$$F(\tau) := \frac{E_2(\tau) - 2E_2(2\tau) - 3E_2(3\tau) + 6E_2(6\tau)}{2\eta^2(\tau)\eta^2(2\tau)\eta^2(3\tau)\eta^2(6\tau)},$$

where $\eta(\tau)$ is the Dedekind eta function and $E_2(\tau)$ is the quasimodular Eisenstein series

$$E_2(\tau) := 1 - 24\sum_{n=1}^{\infty}\sum_{d|n} dq^n,$$

where, as usual, $q = e^{2\pi i \tau}$.

More recently, Yuriy Choliy and the author [CS16] gave the following purely combinatorial multisum representation of $p(n)$:

$$p(n) = \sum_{k=0}^{\lfloor\sqrt{n}\rfloor} D(n,k), \tag{2.114}$$

where $D(n,k)$ is given by the following $(k-1)$-fold sum of terms, each of which is a positive integer:

$$D(n,k) = \sum_{m_k=0}^{U_k} \sum_{m_{k-1}=0}^{U_{k-1}} \cdots \sum_{m_2=0}^{U_2} \left(1 + n - k^2 - \sum_{h=2}^{k} h m_h\right) \prod_{i=2}^{k} (m_i + 1),$$

(2.115)

where

$$U_j := U_j(n,k) = \left\lfloor \frac{n - k^2 - \sum_{h=j+1}^{k} h m_h}{j} \right\rfloor,$$

(2.116)

for $j = 2, 3, 4, \ldots, k$.

Remark 2.66. $D(n,k)$ counts the number of partitions of n with Durfee square of order k.

Remark 2.67. It is shown in [CS16] that the *number of terms* in (2.115) is equal to the the number of first Rogers–Ramanujan partitions, i.e.

$$r_1(n) = \sum_{m_k=0}^{U_k} \sum_{m_{k-1}=0}^{U_{k-1}} \cdots \sum_{m_2=0}^{U_2} 1.$$

2.6.2.2   Rademacher-type Series for Selected Restricted Partition Functions

The circle method can be successfully employed to obtain a convergent series for a variety of restricted partition counting functions analogous to Rademacher's formula for $p(n)$. For illustrative purposes, a few examples are included here:

The following Rademacher-type series for the number of partitions into distinct parts was given by Peter Hagis in 1963 [Hag63]:

$$pd(n) = po(n) = \frac{\pi}{\sqrt{24n+1}} \sum_{\substack{k \geq 1 \\ 2 \nmid k}} \sum_{h \in \mathbb{Z}_k^\times} e^{-2\pi n h / k} \frac{\omega(h,k)}{\omega(2h,k)} I_1\left(\frac{\pi \sqrt{24n+1}}{6k\sqrt{2}}\right).$$

(2.117)

This implies the (weaker) asymptotic formula

$$pd(n) = \frac{18^{1/4}}{(24n+1)^{3/4}} \exp\left(\frac{\pi\sqrt{48n+2}}{12}\right) \left(1 + O(n^{-1/2})\right),$$

(2.118)

as $n \to \infty$.

At some point before World War II, Loo-Keng Hua had discovered a different Rademacher-type series for $pd(n)$. Apparently, he wrote up the paper, and it was accepted for publication in *Acta Arithmetica*, yet it never appeared. So he published it in the *Transactions of the A.M.S.* in 1942 [Hua42].

After completing a PhD at Chicago in 1938 under the supervision of L. E. Dickson, Ivan Niven spent a year at the University of Pennsylvania working with Rademacher under a Harrison Research Fellowship. While at Penn, Rademacher suggested he use the circle method to derive the partition function associated with Schur's mod 6 partition theorem [Niv40]. Recalling that $s(n)$ denotes the number of partitions of n into parts congruent to $\pm 1 \pmod 6$,

$$s(n) = \frac{2\pi}{\sqrt{36n-3}} \sum_{\substack{k \geq 1 \\ 2 \nmid k, 3 \nmid k}} \frac{\tilde{A}_k(n)}{k} I_1\left(\frac{\pi\sqrt{12n-1}}{3k\sqrt{6}}\right)$$

$$+ \frac{12\pi}{\sqrt{36n-3}} \sum_{\substack{k \geq 1 \\ 6 \mid k}} \frac{\tilde{B}_k(n)}{k} I_1\left(\frac{\pi\sqrt{12n-1}}{3k}\right), \quad (2.119)$$

where

$$\tilde{A}_k(n) := \sum_{h \in \mathbb{Z}_k^\times} e^{-2\pi i h/k} \frac{\omega(h,k)\,\omega(6h,k)}{\omega(2h,k)\,\omega(3h,k)}$$

and

$$\tilde{B}_k(n) := \sum_{h \in \mathbb{Z}_k^\times} e^{-2\pi n h/k} \frac{\omega(h,k)\,\omega\left(h,\frac{k}{6}\right)}{\omega\left(h,\frac{k}{2}\right)\,\omega\left(h,\frac{k}{3}\right)}.$$

From the $k = 1$ term of (2.119) together with

$$I_1(z) \sim e^z (2\pi z)^{-1/2},$$

(see [Wat44, p. 203] for details) we may deduce

$$s(n) \sim \frac{6^{1/4}}{(12n-1)^{3/4}} \exp\left(\frac{\pi\sqrt{72n-6}}{18}\right), \quad (2.120)$$

as $n \to \infty$.

For his PhD thesis, Joseph Lehner (a student of Rademacher), derived the analogous formula for the number of partitions arising in the Rogers–Ramanujan identities [Leh41]. Recalling that $r_a(n)$ denotes the number of partitions of n into parts congruent to $\pm a \pmod 5$, for $a = 1$ or 2,

$$r_a(n) = \frac{2\pi}{\sqrt{60n-A}} \sum_{\substack{k \geq 1 \\ 5 \mid k}} \frac{A_k^*(n)}{k} I_1\left(\frac{\pi\sqrt{60n-A}}{15k}\right)$$

$$+ \frac{\pi}{\sqrt{60n-A}} \sum_{\substack{k \geq 1 \\ 5 \nmid k}} \left|\csc\frac{\pi a k}{5}\right| \frac{B_k^*(n)}{k} I_1\left(\frac{\pi\sqrt{60n-A}}{15k}\right), \quad (2.121)$$

where

$$A := A(a) = \begin{cases} 1 & \text{if } a = 1, \\ -11 & \text{if } a = 2, \end{cases}$$

and the $A_k^*(n)$ and $B_k^*(n)$ are Kloosterman sums analogous to $A_k(n)$,

$$A_k^*(n) := \sum_h \omega_a(h,k) \exp\left(-2\pi i \frac{hn}{k}\right), \qquad h \in \mathbb{Z}_k^\times \text{ and } h \equiv \pm a \pmod 5,$$

where in turn

$$\omega_a(h,k) := \exp\left(\pi i \sum_{\mu_a} S\left(\frac{\mu}{k}\right) S\left(\frac{h\mu}{k}\right)\right),$$

with

$$S(x) := \begin{cases} x - \lfloor x \rfloor - \frac{1}{2} & \text{if } x \notin \mathbb{Z}, \\ 0 & \text{if } x \in \mathbb{Z}, \end{cases}$$

and finally, the summation over μ_a means that the summation index runs over those integers strictly between 0 and k that are congruent to $\pm a$ (mod 5),

$$B_k^*(n) := \sum_h \chi_a(h,k) \exp\left(-2\pi i \frac{hn}{k}\right), \qquad h \in \mathbb{Z}_k^\times \text{ and } 5 \nmid k,$$

where

$$\chi_a(h,k) := \exp\left(\pi i \sum_{\mu_a} S\left(\frac{\mu}{5k}\right) S\left(\frac{h\mu}{k}\right)\right).$$

The corresponding asymptotic formulæ are

$$r_1(n) \sim \frac{\sqrt{15 + 3\sqrt{5}}}{(60n - 1)^{3/4}} \exp\left(\frac{\pi\sqrt{60n - 1}}{15}\right) \qquad (2.122)$$

and

$$r_2(n) \sim \frac{\sqrt{15 - 3\sqrt{5}}}{(60n + 11)^{3/4}} \exp\left(\frac{\pi\sqrt{60n + 11}}{15}\right) \qquad (2.123)$$

as $n \to \infty$. Notice that these are slightly more precise variants of the asymptotic formulæ for $r_1(n)$ and $r_2(n)$ deduced from Meinardus' theorem, Eq. (2.110).

Nicolas Smoot [Smo16] derived the analogous Rademacher-type series for $G_1(n)$ (resp. $G_3(n)$), the number of first (resp. second) Göllnitz–Gordon partitions of n:

$$G_a(n) = \frac{\pi\sqrt{2}}{4\sqrt{16n + 4a - 5}} \sum_{\substack{k>0 \\ (k,8)=1}} \left|\csc\left(\frac{\pi a k}{8}\right)\right| \frac{A_{a,1}(n,k)}{k} I_1\left(\frac{\pi\sqrt{16n + 4a - 5}}{8k}\right)$$

$$+ \frac{\pi\sqrt{2}}{\sqrt{16n + 4a - 5}} \sum_{\substack{k>0 \\ (k,8)=4}} \frac{A_{a,4}(n,k)}{k} I_1\left(\frac{\pi\sqrt{16n + 4a - 5}}{4k}\right)$$

$$+ \frac{2\pi}{\sqrt{16n+4a-5}} \sum_{\substack{k>0 \\ (k,8)=8}} \frac{A_{a,8}(n,k)}{k} I_1\left(\frac{\pi\sqrt{16n+4a-5}}{4k}\right), \quad (2.124)$$

where

$$A_{a,d}(n,k) := \sum_{h \in \mathbb{Z}_k^\times} \omega_{a,d}(h,k) e^{-2\pi i n h/k}, \quad (2.125)$$

for $d = 1, 4$, and

$$A_{a,8}(n,k) := \sum_{\substack{h \in \mathbb{Z}_k^\times \\ ah \equiv \pm 1 (\text{mod } 8)}} \omega_{a,8}(h,k) e^{-2\pi i n h/k}, \quad (2.126)$$

with

$$\omega_{a,1}(h,k) = \frac{(-1)^{\lfloor -a(8hH_1+1)/(8k)\rfloor + h - 1}}{\epsilon(1,8)\epsilon(1,4)}$$

$$\times \exp\left\{\frac{\pi i}{4k}[4h(1-a+hH_1(3-4a))-H_1]\right\}$$

$$\omega_{a,4}(h,k) = \frac{i(-1)^{\lfloor ah/4\rfloor}}{\epsilon(4,8)\epsilon(4,4)}$$

$$\times \exp\left\{\frac{\pi i}{4k}[h - H_4 - h(4a-3)(hH_4+1) + a(2hH_4+1)(b-2)]\right\},$$

$$\omega_{a,8}(h,k) = \frac{i(-1)^{\lfloor ah/8\rfloor}}{\epsilon(8,8)\epsilon(8,4)} \exp\left\{\frac{\pi i}{8k}[h(5-4a) - H_8((b-4)^2 - 8)]\right\},$$

where H_8 satisfies

$$hH_8 \equiv -1 \quad (\text{mod } 16k)$$

and H_d satisfies

$$\frac{8hH_d}{d} \equiv -1 \quad (\text{mod } k/d)$$

for $d = 1, 4$. Additionally, b is the least positive residue of ah (mod 8), and for $d = 1, 4$,

$$\epsilon(8,8) := \epsilon\left(h, -8(hH_8+1)/k, k/8, -H_8\right)$$
$$\epsilon(8,4) := \epsilon\left(h, -4(hH_8+1)/k, k/4, -H_8\right)$$
$$\epsilon(d,8) := \epsilon\left(8h/d, -d(8hH_d+1)/k, k/d, -H_d\right)$$
$$\epsilon(d,4) := \epsilon\left(4h/d, -d(8hH_d/d+1), k/d, -2H_d\right)$$

where $\epsilon(a,b,c,d)$ is defined in Equation (1.11).

Chapter 3

Infinite Families ... Everywhere!

3.1 Bailey Chain

The most powerful aspect of Bailey's lemma, the fact that any Bailey pair generates another (see (2.32),(2.33)), and therefore infinitely many, was not noticed by Bailey. This self-replicating aspect of Bailey pairs, known as the *Bailey chain*, was discovered in its full generality by Andrews [And84]. Independently, the iterative potential of Bailey's lemma was observed by Peter Paule [Pau82, Pau85, Pau87].

3.1.1 Iteration of Bailey's Lemma

The most fundamental Bailey pair, the *unit Bailey pair*, in the notation introduced in Section 2.3.3, (see Exercise 2a in Section 2.3.6) is

$$\left(\alpha_n^{(1,1,1)}(x,q), \beta_n^{(1,1,1)}(x,q)\right) = \left(\frac{(-1)^n q^{n(n-1)/2}(1 - xq^{2n})(x;q)_n}{(1-x)(q;q)_n}, \delta_{n,0}\right).$$
(3.1)

Insert (3.1) into (2.32), (2.33) to obtain

$$\alpha_n'(x,q) = \frac{(\rho_1;q)_n(\rho_2;q)_r}{(xq/\rho_1;q)_n(xq/\rho_2;q)_n}\left(\frac{xq}{\rho_1\rho_2}\right)^n \frac{(-1)^n q^{n(n-1)/2}(1 - xq^{2n})(x;q)_n}{(1-x)(q;q)_n}$$
(3.2)

and

$$\beta_n' = \frac{(xq/\rho_1\rho_2;q)_n}{(q;q)_n(xq/\rho_1;q)_n(xq/\rho_2;q)_n}.$$
(3.3)

Take the limit as $\rho_1, \rho_2 \to \infty$ to obtain

$$(\alpha_n'(x,q), \beta_n'(x,q)) = \left(\frac{(-1)^n x^n q^{n(3n-1)/2}(1 - xq^{2n})(x;q)_n}{(1-x)(q;q)_n}, \frac{1}{(q;q)_n}\right)$$

$$= \left(\alpha_n^{(1,1,2)}(x,q), \beta_n^{(1,1,2)}(x,q)\right). \quad (3.4)$$

As we have seen, inserting the Bailey pair (3.4) into (2.39) with $x = 1$ yields the first Rogers–Ramanujan identity.

Let us now repeat the process by inserting (α'_n, β'_n) into (2.32), (2.33). We obtain

$$\alpha''_n(x,q) = \frac{(\rho_1;q)_n(\rho_2;q)_r}{(xq/\rho_1;q)_n(xq/\rho_2;q)_n} \left(\frac{xq}{\rho_1\rho_2}\right)^n$$
$$\times \frac{(-1)^n x^n q^{n(3n-1)/2}(1-xq^{2n})(x;q)_n}{(1-x)(q;q)_n} \quad (3.5)$$

and

$$\beta''_n = \frac{1}{(xq/\rho_1;q)_n(xq/\rho_2;q)_n} \sum_{j=0}^{n} \frac{(\rho_1;q)_j(\rho_2;q)_j(xq/\rho_1\rho_2;q)_{n-j}}{(q;q)_j(q;q)_{n-j}} \left(\frac{xq}{\rho_1\rho_2}\right)^j.$$
$$(3.6)$$

Take the limit as $\rho_1, \rho_2 \to \infty$ to obtain

$$(\alpha''_n(x,q), \beta''_n(x,q))$$
$$= \left(\frac{(-1)^n x^{2n} q^{n(5n-1)/2}(1-xq^{2n})(x;q)_n}{(1-x)(q;q)_n}, \sum_{j=0}^{n} \frac{x^j q^{j^2}}{(q;q)_n(q;q)_{n-j}}\right)$$
$$= \left(\alpha_n^{(1,1,3)}(x,q), \beta_n^{(1,1,3)}(x,q)\right). \quad (3.7)$$

Inserting this Bailey pair into (2.39) with $x = 1$ yields the following Rogers–Ramanujan type identity:

$$\sum_{n=0}^{\infty} \sum_{j=0}^{n} \frac{q^{n^2+j^2}}{(q;q)_j(q;q)_{n-j}} = \frac{f(-q^3,-q^4)}{(q;q)_\infty}, \quad (3.8)$$

or, reversing the order of summation and relabeling the summation indices $j = n_1$ and $n = n_2$,

$$\sum_{n_2 \geq n_1 \geq 0} \frac{q^{n_1^2+n_2^2}}{(q;q)_{n_1}(q;q)_{n_2-n_1}} = \frac{f(-q^3,-q^4)}{(q;q)_\infty}. \quad (3.9)$$

From the associated system of q-difference equations (2.55), one can further deduce

$$\sum_{n_2 \geq n_1 \geq 0} \frac{q^{n_1^2+n_2^2+n_2}}{(q;q)_{n_1}(q;q)_{n_2-n_1}} = \frac{f(-q^2,-q^5)}{(q;q)_\infty} \quad (3.10)$$

and

$$\sum_{n_2 \geq n_1 \geq 0} \frac{q^{n_1^2+n_2^2+n_1+n_2}}{(q;q)_{n_1}(q;q)_{n_2-n_1}} = \frac{f(-q,-q^6)}{(q;q)_\infty}. \quad (3.11)$$

This process of iteration can be repeated indefinitely, and together with the associated system of q-difference equations, this leads to the following theorem of Andrews [And74a], known as the "Andrews–Gordon identity" for reasons to be made clear later.

Theorem 3.1 (Andrews–Gordon identity). *For positive integers k and $1 \leq i \leq k$,*

$$\sum_{n_{k-1} \geq n_{k-2} \geq \cdots \geq n_1 \geq 0} \frac{q^{n_1^2 + n_2^2 + \cdots + n_{k-1}^2 + n_i + n_{i+1} + \cdots + n_{k-1}}}{(q;q)_{n_{k-1} - n_{k-2}} (q;q)_{n_{k-2} - n_{k-3}} \cdots (q;q)_{n_2 - n_1} (q;q)_{n_1}}$$

$$= \frac{f(-q^i, -q^{2k+1-i})}{(q;q)_\infty} = \prod_{\substack{j > 0 \\ j \not\equiv 0, \pm i \pmod{2k+1}}} \frac{1}{1 - q^j}. \quad (3.12)$$

The even modulus analog of (3.12) is due to David Bressoud [Bre80a].

Theorem 3.2 (Bressoud's even modulus analog of the Andrews–Gordon identity). *For positive integers $k > 1$ and $1 \leq i \leq k$,*

$$\sum_{n_{k-1} \geq n_{k-2} \geq \cdots \geq n_1 \geq 0} \frac{q^{n_1^2 + n_2^2 + \cdots + n_{k-1}^2 + n_i + n_{i+1} + \cdots + n_{k-1}}}{(q;q)_{n_{k-1} - n_{k-2}} (q;q)_{n_{k-2} - n_{k-3}} \cdots (q;q)_{n_2 - n_1} (q^2;q^2)_{n_1}}$$

$$= \frac{f(-q^i, -q^{2k-i})}{(q;q)_\infty} = \begin{cases} \displaystyle\prod_{\substack{j \geq 0 \\ j \not\equiv 0, \pm i \pmod{2k}}} \frac{1}{(1 - q^j)} & \text{if } 1 \leq i < k, \\[4mm] (q^k;q^{2k})_\infty (q^k;q^k)_\infty (q;q)_\infty^{-1} & \text{if } i = k. \end{cases} \quad (3.13)$$

The truth of Eq. (3.13) follows from the establishment of the following Bailey pair.

Lemma 3.3. *The pair (α_n, β_n), where*

$$\alpha_n = \alpha_n(x, q) = \frac{(-1)^n (1 - xq^{2n})(x^2; q^2)_n}{(1 - x)(q^2; q^2)_n} \quad (3.14)$$

and

$$\beta_n = \beta_n(x, q) = \frac{1}{(q^2; q^2)_n}, \quad (3.15)$$

forms a Bailey pair with respect to x.

Proof. By Eq. (2.30),

$$\beta_n = \frac{1}{(q;q)_n (xq;q)_n} \sum_{r=0}^{n} \frac{(q^{-n};q)_r}{(xq^{n+1};q)_r} (-1)^r q^{nr - r(r-1)/2} \alpha_n$$

$$= \frac{1}{(q;q)_n (xq;q)_n} \sum_{r=0}^{n} \frac{(q^{-n};q)_r (1 - xq^{2r})(x^2; q^2)_r}{(1 - x)(q^2; q^2)_r (xq^{n+1};q)_r} q^{nr - r(r-1)/2}$$

$$= \frac{1}{(q;q)_n(xq;q)_n} \lim_{\tau \to 0} {}_6\phi_5 \left[\begin{matrix} x, qx^{\frac{1}{2}}, -qx^{\frac{1}{2}}, -x, -xq/\tau, q^{-n} \\ x^{\frac{1}{2}}, -x^{\frac{1}{2}}, -q, -\tau, xq^{n+1} \end{matrix} ; q, \frac{\tau q^n}{x} \right]$$

$$= \frac{1}{(q;q)_n(xq;q)_n} \lim_{\tau \to 0} \frac{(xq;q)_n(\tau/x;q)_n}{(-\tau;q)_n(-q;q)_n} \quad \text{(by Theorem 1.50)}$$

$$= \frac{1}{(q^2;q^2)_n}.$$

$\square$

Sketch of proof of (3.13). Insert the Bailey pair from Lemma (3.3) into Eq. (2.39) to obtain the identity corresponding to the $k = i = 2$ case. Then apply Bailey's Lemma, Eq. (2.34), with $x = 1$ and let $\rho_1, \rho_2 \to \infty$ to obtain the $k = i = 3$ case. Eqs. (2.32) and (2.33) show how to build (α'_n, β'_n). Repeat the process to obtain the $k = i = 4$ case, etc. For a given k, the $i = 1, 2, \ldots, k-1$ subcases may be obtained from the set of difference equations as follows: Define

$$\tilde{Q}_i(x) := \frac{(x^2q^2;q^2)_\infty}{(xq;q)_\infty} J_{(k-1)/2,i/2}(0, x^2, q^2).$$

For $i = 1, 2, \ldots, k$, it can be shown that

$$\tilde{Q}_i(x) = \tilde{Q}_{i-2}(x) + (1+xq)x^{i-2}q^{i-2}\tilde{Q}_{k-i+1}(xq),$$

and these q-difference equations can be used to derive the multisum expressions for a given k and i. The process is straightforward, albeit tedious. $\square$

It is now clear that *every* Rogers–Ramanujan type identity obtained by inserting a Bailey pair into a limiting case of Bailey's lemma is but one member of an infinite family of "multisum = infinite product" identities. This observation was brought to light by Andrews in [And84]; see that paper for additional explicit examples of infinite families of multisum Rogers–Ramanujan type identities.

3.1.2   Exercises

1. Eq. (3.12) is often written in the following form:

$$\sum_{n_1,n_2,\ldots,n_{k-1}\geq 0} \frac{q^{N_1^2+N_2^2+\cdots+N_{k-1}^2+N_i+N_{i+1}+\cdots+N_{k-1}}}{(q;q)_{n_1}(q;q)_{n_2}\cdots(q;q)_{n_{k-1}}}$$

$$= \prod_{\substack{j>0 \\ j\not\equiv 0,\pm i \,(\mathrm{mod}\ 2k+1)}} \frac{1}{1-q^j}, \quad (3.16)$$

where $N_j := n_j + n_{j+1} + \cdots + n_{k-1}$. Show that Eq. (3.16) is in fact equivalent to Eq. (3.12).

2. The identity

$$\sum_{n=0}^{\infty} \frac{q^{n^2}}{(q;q)_{2n}} = \frac{1}{(q;q^2)_{\infty}(q^4;q^{20})_{\infty}(q^8;q^{20})_{\infty}} \tag{3.17}$$

is due to Rogers [Rog94, p. 330 (3), 1st Eq.], and appears twice in Slater's list [Sla52, p. 160, Eq. (79); p. 162, Eq. (98)]. Use the Bailey chain to generalize Eq. (3.17) to

$$\sum_{n_{k-1} \geqq n_{k-2} \geqq \cdots \geqq n_1 \geqq 0} \frac{q^{n_1^2+n_2^2+\cdots+n_{k-1}^2+n_i+n_{i+1}+\cdots+n_{k-1}}}{(q;q)_{n_{k-1}-n_{k-2}}(q;q)_{n_{k-2}-n_{k-3}} \cdots (q;q)_{n_2-n_1}(q;q)_{2n_1}}$$

$$= \prod_{\substack{j>0 \\ j \not\equiv 0,(6k+4),\pm(4k+2),\pm(k+1),\pm(5k+3)(\bmod\ 12k+8)}} \frac{1}{1-q^j} \tag{3.18}$$

3. Derive the identity

$$\sum_{n_1,n_2,\ldots,n_{k-1} \geqq 0} \frac{q^{N_1^2+2N_2^2+2N_3^2+\cdots+2N_{k-1}^2}(-q;q^2)_{N_1}}{(q^2;q^2)_{n_1}(q^2;q^2)_{n_2} \cdots (q^2;q^2)_{n_{k-1}}} = \frac{f(-q^{2k-1},-q^{2k+1})}{\psi(-q)},$$

$$\tag{3.19}$$

where once again $N_j := n_j + n_{j+1} + \cdots + n_{k-1}$.

3.2 Combinatorial Generalizations of Rogers–Ramanujan Type Identities

3.2.1 Gordon's Theorem

In 1961, Basil Gordon published the following combinatorial generalization of the Rogers–Ramanujan identities [Gor61]:

Theorem 3.4 (Gordon). *Let n, k, and i denote integers with $1 \leq i \leq k$. Let $G_{k,i}^*(n)$ denote the number of partitions of n into parts not congruent to $0, \pm i$ (mod $2k+1$). Let $G_{k,i}(n)$ denote the number of partitions $\lambda = (\lambda_1, \lambda_2, \ldots, \lambda_l)$ such that*

- $m_1(\lambda) \leq i-1$,

- $\lambda_j - \lambda_{j+k-1} \geq 2$ *for $j = 1, 2, \ldots, l-k-1$.*

Then $G_{k,i}^*(n) = G_{k,i}(n)$ for all n.

For convenience, we make the following definition.

Definition 3.5. A $(k-i)$-*Gordon partition* of n is a partition λ of the type enumerated by the $G_{k,i}(n)$ of Theorem 3.4. That is, a partition $\lambda = (\lambda_1, \lambda_2, \ldots, \lambda_l)$ in which at most $i-1$ of the parts equal 1, and in which $\lambda_j - \lambda_{j+k-1} \geq 2$ for $j = 1, 2, \ldots, l-k-1$.

Remark 3.6. The first (resp. second) Rogers–Ramanujan identity corresponds to the case $k = i = 2$ (resp. $k = 2$, $i = 1$) of Theorem 3.4.

Remark 3.7. Equation (3.12) is called the "Andrews–Gordon" identity, even though Gordon did not co-author the paper that contains it [And74a], because it is immediately clear that

$$\sum_{n=0}^{\infty} G^*_{k,i}(n)q^n = \frac{f(-q^i, -q^{2k-i+1})}{(q;q)_\infty}, \tag{3.20}$$

and the right hand side of Equation (3.20) is identical to the product side of Equation (3.12). Note, however, that apart from the $k = 2$ case, it is far from obvious that

$$\sum_{n=0}^{\infty} G_{k,i}(n)q^n$$

$$= \sum_{n_{k-1} \geq n_{k-2} \geq \cdots \geq n_1 \geq 0} \frac{q^{n_1^2 + n_2^2 + \cdots + n_{k-1}^2 + n_i + n_{i+1} + \cdots + n_{k-1}}}{(q;q)_{n_{k-1}-n_{k-2}}(q;q)_{n_{k-2}-n_{k-3}} \cdots (q;q)_{n_2-n_1}(q;q)_{n_1}}.$$

Remark 3.8. R. Askey points out in [Ask01] that Andrews independently discovered Theorem 3.4 during his first year on the faculty at Penn State (i.e. 1964–1965) only to learn a few months later that Gordon had in fact already published the result.

Proof of Theorem 3.4. Let k and i denote integers with $1 \leq i \leq k$. Let $G_{k,i}(m,n)$ denote the number of partitions of length m that are enumerated by $G_{k,i}(n)$. Note that the following boundary conditions hold:

$$G_{k,i}(0,0) = 1 \text{ (to account for the empty partition of 0)}, \tag{3.21}$$

$$G_{k,i}(m,n) = 0 \text{ if either } m \leq 0 \text{ or } n \leq 0 \text{ (but } m \text{ and } n \text{ are not both 0)}, \tag{3.22}$$

$$G_{k,0}(m,n) = 0. \tag{3.23}$$

Next we wish to establish the following recurrence:

$$G_{k,i}(m,n) - G_{k,i-1}(m,n) = G_{k,k-i+1}(m-i+1,n-m). \tag{3.24}$$

The left-hand side of (3.24) enumerates those Gordon partitions λ of length m and weight n for which $m_1(\lambda) = i-1$. The right-hand side of (3.24) enumerates those Gordon partitions μ of length $m-i+1$ and weight $n-m$ in which $m_1(\mu) \leq k - i$. We will display a bijection between these two classes of partitions. Let μ denote the partition obtained from λ by subtracting 1 from each of its m parts, noting that μ has length $m - i + 1$, as the $i - 1$ 1s of λ have been reduced to 0s which do not count as parts of a partition. Further, $|\mu| = n - m$. Also note that since $m_1(\lambda) = i - 1$ and $m_1(\lambda) + m_2(\lambda) \leq k - 1$ (because $\lambda_j - \lambda_{j+k-1} \geq 2$), it follows that $m_2(\lambda) \leq (k - 1) - (i - 1) = k - i$, and therefore $m_1(\mu) \leq k - i$. This mapping is easily reversed, and so it is clear that the number of partitions of type μ is $G_{k,k-i+1}(m - i + 1, n - m)$, and therefore the recurrence (3.24) holds.

We next observe that the recurrence (3.24) together with the initial conditions (3.21), (3.22), (3.23) uniquely determine the two-dimensional sequence $G_{k,i}(m, n)$ for all $0 \leq i \leq k$, via a nested induction argument on n and i.

We now recall the double power series $J_{k,i}(0; x; q)$ defined earlier (Def. 2.19).

Define $G_{k,i}^*(m, n)$ implicitly by

$$\sum_{m,n \geq 0} G_{k,i}^*(m, n) x^m q^n := J_{k,i}(0; x; q).$$

Since

$$J_{k,i}(0; 0; q) = J_{k,i}(0; x; 0) = 1,$$

it is immediate that for $1 \leq i \leq k$,

$$G_{k,i}^*(0, 0) = 1 \text{ (to account for the empty partition of 0)}, \tag{3.25}$$

$$G_{k,i}^*(m, n) = 0 \text{ if either } m \leq 0 \text{ or } n \leq 0 \text{ (but } m \text{ and } n \text{ are not both 0)}. \tag{3.26}$$

Also,

$$J_{k,0}(0; x; q) = H_{k,0}(0; xq; q) = 0$$

implies that

$$G_{k,0}^*(m, n) = 0. \tag{3.27}$$

Then, compare coëfficients of $x^m q^n$ on both sides of Lemma 2.21 with $a = 0$ to obtain

$$G_{k,i}^*(m, n) - G_{k,i-1}^*(m, n) = G_{k,k-i+1}^*(m - i + 1, n - m). \tag{3.28}$$

We have thus shown that the $G_{k,i}^*(m, n)$ and $G_{k,i}(m, n)$ satisfy the same recurrence and initial conditions for $0 \leq i \leq k$, and are therefore identical for all m and n.

Further $\sum_{m=0}^{n} G_{k,i}(m, n) = G_{k,i}(n)$, and so

$$\sum_{n=0}^{\infty} G_{k,i}(n) q^n = \sum_{n,m \geq 0} G_{k,i}(m, n) q^n \tag{3.29}$$

$$= J_{k,i}(0; 1; q) \qquad (3.30)$$

$$= \frac{f(-q^i, -q^{2k+1-i})}{(q; q)_\infty} \qquad (3.31)$$

$$= \sum_{n=0}^{\infty} G_{k,i}^*(n) q^n. \qquad (3.32)$$

Therefore we have $G_{k,i}^*(n) = G_{k,i}(n)$ for all n. □

3.2.2 Even Modulus Analog of Gordon's Theorem

Theorem 3.4 can be extended in various ways. For instance, Bressoud [Bre79] found a complementary result involving partitions in residue classes with even moduli:

Theorem 3.9 (Bressoud). *Let k be a fixed positive integer and let $1 \leq i < k$. Let $B_{k,i}^*(n)$ denote the number of partitions into parts not congruent to $0, \pm i \pmod{2k}$. Let $B_{k,i}(n)$ denote the number of (k, i)-Gordon partitions $\lambda = (\lambda_1, \ldots, \lambda_l)$ of n with the property that*

$$\lambda_j - \lambda_{j+k-2} \leq 1 \implies \sum_{h=0}^{k-2} \lambda_{j+h} \equiv (i-1) \pmod 2. \qquad (3.33)$$

Then $B_{k,i}(n) = B_{k,i}^*(n)$ for all n.

Remark 3.10. The method of proof of Theorem 3.9 is completely analogous to that of Theorem 3.4. The key observation is that

$$(-q; q)_\infty J_{(k-1)/2, i/2}(0; 1; q^2) = \frac{f(-q^i, -q^{2k-i})}{(q; q)_\infty}.$$

Remark 3.11. The series-product identity corresponding to Theorem 3.9 is Equation (3.13).

Remark 3.12. It was observed by Shashank Kanade and Matthew Russell [KLRS17] that an important role is also played by (k, i)-Gordon partitions with the additional restriction that they satisfy the *opposite* parity condition of (3.33). The corresponding series were named "ghost series."

Next, we state Andrews' generalization [And67] of the Göllnitz–Gordon identities, Theorems 2.42 and 2.43. The first and second Göllnitz–Gordon identities correspond respectively to the $k = i = 2$ and $k = i + 1 = 2$ cases of the following theorem.

Theorem 3.13. *Let n, k, and i denote integers with $1 \leq i \leq i$. Let $B_{k,i}^*$ denote the number of partitions of n into parts neither congruent to 2 (mod 4) nor to $0, \pm(2i-1)$ (mod 4k). Let $B_{k,i}(n)$ denote the number of partitions $\lambda = (\lambda_1, \lambda_2, \ldots, \lambda_l)$ (l arbitrary) of n such that*

- $m_j > 1$ *only if j is even,*

- $\lambda_j - \lambda_{j+k-1} \geqq 2$ *for all $1 \leqq j \leqq l(\lambda) - k + 1$,*

- $\lambda_j - \lambda_{j+k-1} = 2$ *only if λ_j is odd,*

- $m_1(\lambda) + m_2(\lambda) \leqq i - 1$.

Then $B_{k,i}(n) = B^*_{k,i}(n)$ for all n.

Proof. The proof is essentially identical to that of Theorem 3.4, with $J_{k,i}(1/q; x; q^2)$ used in place of $J_{k,i}(0; x; q)$, and is therefore left as an exercise. For a full treatment, see [And67] or [And76, Theorem 7.11, pp. 114–115]. □

In the 1997 inaugural issue of the *Ramanujan Journal*, Andrews and Plínio Santos presented another infinite family Rogers–Ramanujan type partition theorem in the same spirit as Theorems 3.4, 3.9, and 3.13 [AS97].

3.2.3 Partitions with Attached Odd Parts

Theorem 3.14 (Andrews and Santos). *Let n, k and i denote integers with $1 \leq i \leq k$. Let $AS^*_{k,i}(n)$ denote the number of partitions of n into parts that are either even but not congruent to $0, \pm(4i-2), 4k \pmod{8k}$ or odd and congruent to $\pm(2i-1) \pmod{4k}$. Let $AS^{**}_{k,i}(n)$ denote the number of partitions of n into parts that are either even but not a multiple of $4k$, or distinct, odd and congruent to $\pm(2i-1) \pmod{4k}$. Let $AS_{k,i}(n)$ denote the number of partitions λ of n in which*

- $m_2(\lambda) \leqq i - 1$,

- $m_{2j}(\lambda) + m_{2j+2}(\lambda) \leqq k - 1$,

- $m_{2j+1}(\lambda) > 0$ *only if $m_{2j}(\lambda) + m_{2j+2}(\lambda) = k - 1$, where we follow the convention that $m_0(\lambda) = k - i$ in the $j = 0$ case.*

Then $AS^*_{k,i}(n) = AS^{**}_{k,i}(n) = AS_{k,i}(n)$ for all n.

Remark 3.15. The $k = 2$ case of Theorem 3.14 gives a natural combinatorial interpretation of the Jackson–Slater identities (2.14) and (2.15).

Remark 3.16. Andrews' AMS *Memoir* "On the general Rogers–Ramanujan theorem" [And74b] presents a very general result that includes Gordon's theorem, the Göllnitz–Gordon identities, and Schur's mod 6 partition theorem as special cases. The author's paper [Sil07] shows how dilated versions of Theorems 3.4, 3.9, and 3.14 naturally arise as combinatorial interpretations of q-identities obtained via the general multiparameter Bailey pair given in (2.43), (2.44) and two other related multiparameter Bailey pairs.

3.2.4   Exercises

1. Prove Theorem 3.9.

2. Prove Theorem 3.14.

Chapter 4

From Infinite to Finite

"The progress of mathematics may be viewed as a movement from the infinite to the finite."—Gian-Carlo Rota [Rot09, p. 214]

4.1 Andrews' Method of q-Difference Equations

It is common to let $G(q)$ and $H(q)$ denote the series sides of the first and second Rogers–Ramanujan identities (2.3, 2.4) respectively, i.e.

$$G(q) := \sum_{j=0}^{\infty} \frac{q^{j^2}}{(q;q)_j}, \qquad H(q) := \sum_{j=0}^{\infty} \frac{q^{j(j+1)}}{(q;q)_j}. \tag{4.1}$$

4.1.1 Andrews' "t-Generalization"

Let us now follow Andrews [And86, Chapter 9] and define the following generalization of $G(q)$:

$$\mathfrak{G}(t) := \mathfrak{G}(t,q) := \sum_{j=0}^{\infty} \frac{t^{2j} q^{j^2}}{(t;q)_{j+1}}. \tag{4.2}$$

Notice that $\mathfrak{G}(t)$ is *not* the generalization of $G(q)$ used to prove the first Rogers–Ramanujan identity in Chapter 2; here the exponent on t is $2j$ rather than j, and there appears to be an "extra" factor $(1-t)$ in the denominator of the general term. It so happens that $\mathfrak{G}(t)$ has some interesting properties, which we shall now explore.

Immediately we observe that

$$\lim_{t \to 1^-} (1-t)\mathfrak{G}(t) = G(q),$$

and

$$\lim_{t \to -1^-} (1-t)\mathfrak{G}(t) = f_0(q),$$

which is Ramanujan's fifth order mock theta function $f_0(q)$, Eq. (2.98).

Next we derive a first order nonhomogeneous q-difference equation satisfied by $\mathfrak{G}(t)$ as follows:

$$\mathfrak{G}(t) = \sum_{j=0}^{\infty} \frac{t^{2j}q^{j^2}}{(t;q)_{j+1}}$$

$$= \frac{1}{1-t} + \sum_{j=1}^{\infty} \frac{t^{2j}q^{j^2}}{(t;q)_{j+1}}$$

$$= \frac{1}{1-t} + \sum_{j=0}^{\infty} \frac{t^{2j+2}q^{j^2+2j+1}}{(t;q)_{j+2}}$$

$$= \frac{1}{1-t} + \frac{t^2q}{1-t}\mathfrak{G}(tq).$$

Recording the extremes, and multiplying both sides by $(1-t)$, we observe

$$(1-t)\mathfrak{G}(t) = 1 + t^2q\mathfrak{G}(tq),$$

or equivalently

$$\mathfrak{G}(t) = 1 + t\mathfrak{G}(t) + t^2q\mathfrak{G}(tq). \tag{4.3}$$

4.1.2   A q-Fibonacci Sequence and its Fermionic Representation

Thinking of $\mathfrak{G}(t)$ as a power series in t with coëfficients depending on q, let us write

$$\mathfrak{G}(t) = \sum_{n=0}^{\infty} D_n(q)t^n,$$

or rather, taking into account (4.3),

$$\sum_{n=0}^{\infty} D_n(q)t^n = 1 + \sum_{n=0}^{\infty} D_n(q)t^{n+1} + \sum_{n=0}^{\infty} D_n(q)(tq)^n t^2 q,$$

which is equivalent to

$$\sum_{n=0}^{\infty} D_n(q)t^n = 1 + \sum_{n=1}^{\infty} D_{n-1}(q)t^n + \sum_{n=2}^{\infty} D_{n-2}(q)q^{n-1}t^n. \tag{4.4}$$

Hence by comparing coëfficients of t^n on either side of (4.4), we find

$$D_0(q) = D_1(q) = 1, \tag{4.5}$$

and

$$D_n(q) = D_{n-1}(q) + q^{n-1}D_{n-2}(q), \tag{4.6}$$

when $n > 1$, and so the $D_n(q)$ are polynomials. Not just any polynomials, though! Notice that setting $q = 1$ in the sequence of $D_n(q)$ gives the Fibonacci sequence i.e. the $D_n(q)$ are q-analogs of the Fibonacci numbers. If we follow the common indexing convention for the Fibonacci numbers where $F_1 = F_2 = 1$ and $F_n = F_{n-1} + F_{n-2}$ when $n > 2$, we find that

$$D_n(1) = F_{n+1}$$

for $n \geq 0$.

Further, we can derive a formula for $D_n(q)$ from the information we already have.

$$\sum_{n=0}^{\infty} D_n(q)t^n = \mathfrak{G}(t)$$

$$= \sum_{j=0}^{\infty} \frac{t^{2j} q^{j^2}}{(t; q)_{j+1}}$$

$$= \sum_{j=0}^{\infty} t^{2j} q^{j^2} \sum_{k=0}^{\infty} \binom{j+k}{k}_q t^k \quad \text{(by (1.80))}$$

$$= \sum_{j=0}^{\infty} \sum_{k=0}^{\infty} q^{j^2} \binom{j+k}{k}_q t^{2j+k}$$

$$= \sum_{n=0}^{\infty} t^n \sum_{j=0}^{\infty} q^{j^2} \binom{n-j}{j}_q \quad \text{(by letting } n = 2j + k\text{).}$$

Comparing coëfficients of t^n in the extremes, we find

$$D_n(q) = \sum_{j \geq 0} q^{j^2} \binom{n-j}{j}_q. \tag{4.7}$$

However, all but finitely many terms on the right-hand side of (4.7) are zero, so we may instead write

$$D_n(q) = \sum_{j=0}^{\lfloor n/2 \rfloor} q^{j^2} \binom{n-j}{j}_q. \tag{4.8}$$

Of course, we get as an immediate bonus, an expression for the nth Fibonacci number as a sum of binomial coëfficients:

$$F_n = D_{n-1}(1) = \sum_{j=0}^{\lfloor (n-1)/2 \rfloor} \binom{n-j-1}{j}.$$

Notice that

$$\lim_{n \to \infty} D_n(q) = \lim_{n \to \infty} \sum_{j=0}^{\lfloor n/2 \rfloor} \frac{q^{j^2} (q; q)_{n-j}}{(q; q)_j (q; q)_{n-2j}} = \sum_{j=0}^{\infty} \frac{q^{j^2}}{(q; q)_j} = G(q),$$

showing that the $D_n(q)$ are a sequence of polynomials that converge to the power series expansion of $G(q)$.

Physicists call the representation of $D_n(q)$ given in (4.8) the *fermionic* form.

4.1.3 Searching for a Bosonic Form

The *bosonic* representation of $D_n(q)$ is as follows:

$$d_n(q) := \sum_{j \in \mathbb{Z}} (-1)^j q^{j(5j+1)/2} \binom{n}{\lfloor \frac{n+5j+1}{2} \rfloor}_q. \qquad (4.9)$$

To see how we might reasonably guess (4.9), consider the following:

$$D_0(q) = 1 = \binom{1}{0}_q$$

$$D_2(q) = 1 + q = \binom{2}{1}_q$$

$$D_4(q) = 1 + q + q^2 + q^3 + q^4 = \binom{4}{2}_q - q^2 \binom{4}{4}_q$$

$$D_6(q) = 1 + q + q^2 + q^3 + 2q^4 + 2q^5 + 2q^6 + q^7 + q^8 + q^9$$
$$= \binom{6}{3}_q - q^2 \binom{6}{5}_q - q^3 \binom{6}{6}_q$$

$$D_8(q) = \binom{8}{4}_q - q^2 \binom{8}{6}_q - q^3 \binom{8}{7}_q$$

$$D_{10}(q) = \binom{10}{5}_q - q^2 \binom{10}{7}_q - q^3 \binom{10}{8}_q + (q^9 + q^{11}) \binom{10}{10}_q.$$

The above list, using the symmetry of the q-binomial coëfficients, can be written as follows:

$$D_0(q) = \binom{1}{0}_q$$

$$D_2(q) = \binom{2}{1}_q$$

$$D_4(q) = \binom{4}{2}_q - q^2 \binom{4}{0}_q$$

$$D_6(q) = \binom{6}{3}_q - q^2 \binom{6}{1}_q - q^3 \binom{6}{6}_q$$

$$D_8(q) = \binom{8}{4}_q - q^2 \binom{8}{2}_q - q^3 \binom{8}{7}_q$$

$$D_{10}(q) = \binom{10}{5}_q - q^2\binom{10}{3}_q - q^3\binom{10}{8}_q + q^9\binom{10}{0}_q + q^{11}\binom{10}{10}_q.$$

From here, it appears reasonable to guess that

$$D_{2m}(q) = \sum_{j \in \mathbb{Z}} (-1)^j q^{j(5j+1)/2} \binom{2m}{\lfloor m + \frac{5j+1}{2} \rfloor}_q. \tag{4.10}$$

A similar investigation of the odd-subscripted Ds leads one to guess that

$$D_{2m+1}(q) = \sum_{j \in \mathbb{Z}} (-1)^j q^{j(5j+1)/2} \binom{2m+1}{\lfloor m + 1 + \frac{5j}{2} \rfloor}_q. \tag{4.11}$$

Combining (4.10) and (4.11) gives (4.9).

For additional discussion on conjecturing (4.9) with the aid of a computer algebra system, see [AB90] or [Sil04b].

Once conjectured, to prove that $D_n(q) = d_n(q)$, one can verify that the $d_n(q)$ satisfy the same recurrence and initial conditions as the $D_n(q)$ i.e. $d_n(q)$ satisfies (4.4) and (4.5).

The recurrence and initial conditions can be verified either "by hand" or with the aid of a specialized computer algebra package such as the qZeil package for *Mathematica* offered by the Research Institute for Symbolic Computation (RISC) in Linz, Austria; see their website

http://www.risc.jku.at/research/combinat/software/.

Either way, the presence of the floor function in $d_n(q)$ presents a technical difficulty of sorts, which is best dealt with by considering even and odd cases separately:

$$d_{2m}(q) = \sum_{k \in \mathbb{Z}} \left(q^{10k^2+k} \binom{2m}{m+5k}_q - q^{10k^2-9k+2} \binom{2m}{m+5k-2}_q \right), \tag{4.12}$$

$$d_{2m+1}(q) = \sum_{k \in \mathbb{Z}} \left(q^{10k^2+k} \binom{2m+1}{m+5k+1}_q - q^{10k^2-9k+2} \binom{2m+1}{m+5k-2}_q \right). \tag{4.13}$$

The recurrence (4.4) can be written in the form

$$D_{n+2}(q) - D_{n+1}(q) - q^{n+1} D_n = 0.$$

So let us suppose that $n = 2m$, and consider that by judicious applications of the q-Pascal triangle recurrences (1.50) and (1.51) to terms as appropriate, we will find that

$$d_{2m+2}(q) - d_{2m+1}(q) - q^{2m+1} d_{2m}$$

$$= \sum_{k \in \mathbb{Z}} \left(q^{10k^2+k} \binom{2m+2}{m+5k+1}_q - q^{10k^2-9k+2} \binom{2m+2}{m+5k-1}_q \right.$$

$$- q^{10k^2+k} \binom{2m+1}{m+5k+1}_q + q^{10k^2-9k+2} \binom{2m+1}{m+5k-2}_q$$

$$\left. -q^{2m+1+10k^2+k} \binom{2m}{m+5k}_q + q^{2m+3+10k^2-9k} \binom{2m}{m+5k-2}_q \right)$$

(which by application of (1.50) to the first term
and (1.51) to the second term)

$$= \sum_{k \in \mathbb{Z}} \left(q^{10k^2-4k+m+1} \binom{2m+1}{m+5k}_q - q^{10k^2-4k+m+1} \binom{2m+1}{m+5k-1}_q \right.$$

$$\left. -q^{2m+1+10k^2+k} \binom{2m}{m+5k}_q + q^{2m+3+10k^2-9k} \binom{2m}{m+5k-2}_q \right)$$

(which by application of (1.51) to the first term
and (1.50) to the second term)

$$= 0.$$

And thus $d_{n+2}(q) - d_{n+1}(q) - q^{n+1}d_n = 0$ when n is even. A similar calculation will demonstrate that $d_{n+2}(q) - d_{n+1}(q) - q^{n+1}d_n = 0$ when n is odd:

$$d_{2m+3}(q) - d_{2m+2}(q) - q^{2m+2}d_{2m+1}$$

$$= \sum_{k \in \mathbb{Z}} \left(q^{10k^2+k} \binom{2m+3}{m+5k+2}_q - q^{10k^2-9k+2} \binom{2m+3}{m+5k-1}_q \right.$$

$$- q^{10k^2+k} \binom{2m+2}{m+5k+1}_q + q^{10k^2-9k+2} \binom{2m+2}{m+5k-1}_q$$

$$\left. -q^{2m+2+10k^2+k} \binom{2m+1}{m+5k+1}_q + q^{2m+4+10k^2-9k} \binom{2m+1}{m+5k-2}_q \right)$$

(which, by application of (1.51) to the first term
and (1.50) to the second term)

$$= \sum_{k \in \mathbb{Z}} \left(-q^{10k^2+k+2m+2} \binom{2m+1}{m+5k+1}_q + q^{10k^2-9k+2m+4} \binom{2m+2}{m+5k-2}_q \right.$$

$$\left. +q^{10k^2+6k+m+2} \binom{2m+2}{m+5k+2}_q - q^{10k^2-14k+m+6} \binom{2m+2}{m+5k-2}_q \right)$$

(which, by application of (1.50) to the third term
and (1.51) to the last term)

$$= q^{10k^2+6k+m+2} \binom{2m+1}{m+5k+2}_q - q^{10k^2-14k+m+6} \binom{2m+1}{m+5k-3}_q$$

$$= 0,$$

where the last equality follows by replacing k by $k+1$ in the last term.

It is easy to verify that $d_0(q) = d_1(q) = 1$, and so we have proved the following finite analog of (2.3):

$$\sum_{j=0}^{\lfloor n/2 \rfloor} q^{j^2} \binom{n-j}{j}_q = \sum_{j \in \mathbb{Z}} (-1)^j q^{j(5j+1)/2} \binom{n}{\lfloor \frac{n+5j+1}{2} \rfloor}_q. \tag{4.14}$$

Upon letting $n \to \infty$ in (4.14), and applying Jacobi's triple product identity to the right-hand side, the first Rogers–Ramanujan identity (2.3) is recovered.

Of course, we also now have another formula for the Fibonacci sequence:

$$F_{n+1} = \sum_{j \in \mathbb{Z}} (-1)^j \binom{n}{\lfloor \frac{n+5j+1}{2} \rfloor}. \tag{4.15}$$

Historical Note 4.1. The polynomial sequence $D_n(q)$ and its representation on the left side of (4.14) were known to MacMahon, while the same sequence represented by the right side of (4.14) was used by Schur in his proof of the Rogers–Ramanujan identities. Accordingly, it seems appropriate to refer to (4.14) is the *MacMahon–Schur polynomial analog of the first Rogers–Ramanujan identity.* The process of finding a polynomial analog of a series-product identity may be referred to as *finitization*; this term appears to have first been used by Ezer Melzer [Mel94].

4.1.4 MacMahon–Schur Polynomial Analog of the Second Rogers–Ramanujan Identity

Analogously, one can derive the MacMahon–Schur polynomial analog of the second Rogers–Ramanujan identity,

$$E_n(q) := \sum_{j=0}^{\lfloor n/2 \rfloor} q^{j(j+1)} \binom{n-j}{j}_q = \sum_{j \in \mathbb{Z}} (-1)^j q^{j(5j+3)/2} \binom{n+1}{\lfloor \frac{n+5j+3}{2} \rfloor}_q. \tag{4.16}$$

The $E_n(q)$ are also a q-Fibonacci sequence, but different from the $D_n(q)$; they satisfy $E_0(q) = E_1(q) = 1$ and

$$E_n(q) = E_{n-1}(q) + q^n E_{n-2}(q) \tag{4.17}$$

when $n \geq 2$.

4.1.5   Finite Rogers–Ramanujan using q-Trinomial Coëfficients

Andrews [And90b] found that $D_n(q)$ and $E_n(q)$ can be represented in terms of q-trinomial coëfficients:

$$D_n(q) = \sum_{k\in\mathbb{Z}} q^{10k^2+k} \left(\!\!\binom{n}{5k}\!\!\right)_q - \sum_{k\in\mathbb{Z}} q^{10k^2-9k+2} \left(\!\!\binom{n}{5k-2}\!\!\right)_q \qquad (4.18)$$

$$E_n(q) = \sum_{k\in\mathbb{Z}} q^{10k^2+3k} \left(\!\!\binom{n}{5k+1}\!\!\right)_q - \sum_{k\in\mathbb{Z}} q^{10k^2-7k+1} \left(\!\!\binom{n}{5k-2}\!\!\right)_q. \qquad (4.19)$$

For more on q-Fibonacci polynomials, see the paper by Johann Cigler [Cig04].

4.1.6   Exercises

1. Let $\rho(j)$ be defined by

$$D_n(q) = \sum_{j\geq 0} \rho(j)q^j.$$

   Show that $\rho(j)$ counts the number of partitions of j into parts less than n that mutually differ by at least 2.

2. Derive Eq. (4.16).

3. Find a partition theoretic interpretation of $E_n(q)$.

4.2   Reciprocal Duality

The *dual* of the polynomial

$$P(q) = a_0 + a_1 q + a_2 q^2 + \cdots + a_m q^m,$$

where $a_m \neq 0$, is

$$q^m P\left(\frac{1}{q}\right) = a_m + a_{m-1}q + a_{m-2}q^2 + \cdots + a_0 q^m.$$

Let us consider separately the subsequences $D_{2n}(q)$ and $D_{2n+1}(q)$ and find the corresponding identities dual to them.

First, the case $n = 2m$ of (4.14) is

$$\sum_{j=0}^{m} q^{j^2} \binom{2m-j}{j}_q$$

$$= \sum_{k \in \mathbb{Z}} \left(q^{10k^2+k} \binom{2m}{m+5k}_q - q^{10k^2-9k+2} \binom{2m}{m+5k-2}_q \right). \quad (4.20)$$

One can prove (e.g. by induction) that $\deg(D_{2m}(q)) = m^2$, so the dual polynomials are $q^{m^2} D_{2m}(1/q)$ and are given by

$$q^{m^2} D_{2m}(1/q) = \sum_{j=0}^{m} q^{m^2-j^2} \binom{2m-j}{j}_{1/q}$$

$$= \sum_{k \in \mathbb{Z}} \left(q^{m^2-10k^2-k} \binom{2m}{m+5k}_{1/q} - q^{m^2-10k^2+9k-2} \binom{2m}{m+5k-2}_{1/q} \right),$$
$$(4.21)$$

or, after application of Eq. (1.53),

$$q^{m^2} D_{2m}(1/q) = \sum_{j=0}^{m} q^{(m-j)^2} \binom{2m-j}{j}_q$$

$$= \sum_{k \in \mathbb{Z}} \left(q^{15k^2-k} \binom{2m}{m+5k}_q - q^{15k^2-11k+2} \binom{2m}{m+5k-2}_q \right), \quad (4.22)$$

and finally, after reindexing the fermionic sum, we obtain the identity

$$\sum_{j=0}^{m} q^{j^2} \binom{m+j}{2j}_q$$

$$= \sum_{k \in \mathbb{Z}} \left(q^{15k^2-k} \binom{2m}{m+5k}_q - q^{15k^2-11k+2} \binom{2m}{m+5k-2}_q \right), \quad (4.23)$$

an identity due to Andrews [And86, p. 80, Eq. 8.42].

If, in (4.23), m is sent to ∞, we obtain:

$$\sum_{j=0}^{\infty} \frac{q^{j^2}}{(q;q)_{2j}} = \frac{1}{(q;q)_\infty} \sum_{k \in \mathbb{Z}} \left(q^{15k^2-k} - q^{15k^2-11k+2} \right)$$

$$= \frac{f(q,q^{14}) - q^2 f(q^4,q^{11})}{(q;q)_\infty} \quad \text{(by (1.58))}$$

$$= \frac{(q^2,q^8,q^{10};q^{10})_\infty (q^6,q^{14};q^{20})_\infty}{(q;q)_\infty} \quad \text{(by (1.62))}$$

$$= \frac{f(-q^8,-q^{12})}{\psi(-q)}$$

$$= \frac{G(q^4)}{(q;q^2)_\infty}.$$

The identity

$$\sum_{j=0}^{\infty} \frac{q^{j^2}}{(q;q)_{2j}} = \frac{G(q^4)}{(q;q^2)_{\infty}} \tag{4.24}$$

is due to Rogers [Rog94, p. 330 (3), 1st Eq.]; cf. [Sla52, p. 160, Eq. (79)].

A similar development leads to the following identities:

$$q^{m(m+1)} D_{2m+1}(q^{-1}) = \sum_{j=0}^{m} q^{j(j+1)} \binom{m+j+1}{2j+1}_q$$

$$= \sum_{k \in \mathbb{Z}} \left(q^{15k^2+4k} \binom{2m+1}{m+5k+1}_q - q^{15k^2+14k+3} \binom{2m+1}{m+5k+3}_q \right) \tag{4.25}$$

[Sil03, p. 46, Eq. (3.94-b)];

$$q^{m^2} E_{2m}(q^{-1}) = \sum_{j=0}^{m} q^{j(j+1)} \binom{m+j}{2j}_q$$

$$= \sum_{k \in \mathbb{Z}} \left(q^{15k^2+2k} \binom{2m+1}{m+5k+1}_q - q^{15k^2+8k+1} \binom{2m+1}{m+5k+2}_q \right) \tag{4.26}$$

[Sil03, p. 47, Eq. (3.99-b)];

$$q^{m^2} E_{2m+1}(q^{-1}) = \sum_{j=0}^{m} q^{j(j+2)} \binom{m+j+1}{2j+1}_q$$

$$= \sum_{k \in \mathbb{Z}} \left(q^{15k^2+7k} \binom{2m+2}{m+5k+2}_q - q^{15k^2+13k+2} \binom{2m+2}{m+5k+3}_q \right) \tag{4.27}$$

[Sil03, p. 47, Eq. (3.96-b)]. Sending $m \to \infty$ in each of the preceding three identities yields, respectively, the following identities of Rogers [Rog17, pp. 331–332], cf. [Sla52, p. 162, Eqs. (94), (99), (96), respectively]:

$$\sum_{j=0}^{\infty} \frac{q^{j(j+1)}}{(q;q)_{2j+1}} = \frac{H(-q)}{(q;q^2)_{\infty}}, \tag{4.28}$$

$$\sum_{j=0}^{\infty} \frac{q^{j(j+1)}}{(q;q)_{2j}} = \frac{G(-q)}{(q;q^2)_{\infty}}, \tag{4.29}$$

$$\sum_{j=0}^{\infty} \frac{q^{j(j+2)}}{(q;q)_{2j+1}} = \frac{H(q^4)}{(q;q^2)_{\infty}}. \tag{4.30}$$

4.3 Bressoud Polynomials

Let us now alter the function $\mathfrak{G}(t)$ slightly:

$$\tilde{\mathfrak{G}}(t) := \tilde{\mathfrak{G}}(t,q) := \sum_{j=0}^{\infty} \frac{t^j q^{j^2}}{(t;q)_{j+1}}. \tag{4.31}$$

While we retain the property

$$\lim_{t \to 1^-} (1-t)\tilde{\mathfrak{G}}(t) = G(q),$$

unlike $\mathfrak{G}(t)$, we do not have a first order nonhomogeneous q-difference equation satisfied by $\tilde{\mathfrak{G}}(t)$. Nonetheless, we can proceed as with $\mathfrak{G}(t)$ to derive a formula for the coëfficents of t^n in the power series expansion:

$$\sum_{n=0}^{\infty} B_n(q)t^n = \tilde{\mathfrak{G}}(t)$$

$$= \sum_{j=0}^{\infty} \frac{t^j q^{j^2}}{(t;q)_{j+1}}$$

$$= \sum_{j=0}^{\infty} t^j q^{j^2} \sum_{k=0}^{\infty} \binom{j+k}{k}_q t^k$$

$$= \sum_{j=0}^{\infty} \sum_{k=0}^{\infty} q^{j^2} \binom{j+k}{k}_q t^{j+k}$$

$$= \sum_{n=0}^{\infty} t^n \sum_{j=0}^{\infty} q^{j^2} \binom{n}{j}_q \quad \text{(by letting } n = j+k\text{)}.$$

Comparing coëfficients of t^n in the extremes, we find

$$B_n(q) = \sum_{j=0}^{n} q^{j^2} \binom{n}{j}_q. \tag{4.32}$$

Thus we see that the $B_n(q)$ are polynomials in q, but since we do not have a q-difference equation satisfied by the $\tilde{\mathfrak{G}}(t)$, we do not (yet!) have a recurrence satisfied by the $B_n(q)$.

So let us turn to the q-Zeilberger algorithm to find a recurrence satisfied by the $B_n(q)$. Using the qEKHAD Maple package written by D. Zeilberger and available for free download from his website, we find that $B_n(q)$ is annihilated by the operator

$$\mathfrak{A}(\eta) = \eta^2 - (q^{2n+3} + q - q^{n+2} + 1)\eta + q(1 - q^{n+1}). \tag{4.33}$$

To prove this, we let

$$F_q(n, k) = q^{k^2} \binom{n}{k}_q$$

and cleverly construct

$$G_q(n, k) := \frac{q^{2n+k+3}(1 - q^{n+1})}{q^{n+1} - q^k} q^{k^2} \binom{n}{k}_q$$

with the motive that

$$F_q(n + 2, k) - (q^{2n+3} + q - q^{n+2} + 1)F_q(n + 1, k) + q(1 - q^{n+1})$$
$$= G_q(n, k) - G_q(n, k - 1), \quad (4.34)$$

an identity of rational functions that is routine to check. Sum (4.34) over k to establish that

$$B_{n+2}(q) = (q^{2n+3} + q - q^{n+2} + 1)B_{n+1}(q) - q(1 - q^{n+1})B_n(q). \quad (4.35)$$

Also, we have

$$B_0(q) = 1 \text{ and } B_1(q) = 1 + q. \quad (4.36)$$

The $B_n(q)$ are a sequence of polynomials that converges to the first Rogers–Ramanujan function in the sense that

$$\lim_{n \to \infty} B_n(q) = G(q).$$

For brevity, we may write $B_\infty(q) = G(q)$. However, unlike the $D_n(q)$ (for which $D_\infty(q)$ also equals $G(q)$), the $B_n(q)$ are *not* q-analogs of the Fibonacci numbers. Observe:

$$B_{n+2}(1) = 2B_{n+1}(1)$$

with $B_0(1) = 1$ and $B_1(1) = 2$, so the $B_n(q)$ is a q-analog of the sequence $\{2^n\}_{n=0}^\infty$.

In [Bre81, p. 211, Eq. (1.1)], David Bressoud gave the following identity for the $B_n(q)$:

$$\sum_{k=0}^n q^{k^2} \binom{n}{k}_q = \sum_{k=-\infty}^\infty (-1)^k q^{k(5k+1)/2} \binom{2n}{n + 2k}_q. \quad (4.37)$$

We have already established that the left side of (4.37), $B_n(q)$, satisfies a certain recurrence (4.34) and initial conditions (4.36). Let $B_n^*(q)$ denote the right side of (4.37). If we can establish that $B_n^*(q)$ satisfies the same recurrence and initial conditions as $B_n(q)$, we will have proved Bressoud's identity (4.37). The initial conditions $B_0^*(q) = 1$ and $B_1^*(q) = 1 + q$ are easily verified. We turn to our favorite implementation of the q-Zeilberger algorithm, expecting to find that the $B_n^*(q)$ satisfy the recurrence (4.35). But, alas, that is *not* what the algorithm finds! Instead, the algorithm finds that the $B_n^*(q)$ satisfies the following *fifth* order recurrence:

$$B_{n+5}^*(q) - (1 + q + q^2 + q^3 + q^4 + q^{2n+7} + q^{2n+8} + q^{2n+9})B_{n+4}^*(q)$$

$$+ q \left(q^{4(n+3)} + q^{2(n+5)} + q^{2n+6} + q^{2n+7} + 2q^{2n+8} + 2q^{2n+9} \right.$$

$$\left. + q^{4n+13} + q^{4n+14} + q^6 + q^5 + 2q^4 + 2q^3 + 2q^2 + q + 1 \right) B_{n+3}^*(q)$$

$$- q^3 \left(q^{4(n+3)} + q^{2n+7} + q^{2n+8} + q^{2n+9} + q^{4n+11} + q^{6n+15} \right.$$

$$\left. + q^6 + q^5 + 2q^4 + 2q^3 + 2q^2 + q + 1 \right) B_{n+2}^*(q)$$

$$- q^6 \left(q^{5(n+2)} + q^{2n+3} + q^{2n+4} + q^{2n+5} + q^{5n+8} \right.$$

$$\left. - q^4 - q^3 - q^2 - q - 1 \right) B_{n+1}^*(q)$$

$$- q^{10} \left(q^{2n+1} - 1 \right) \left(q^{2n+2} - 1 \right) B_n^*(q), \quad (4.38)$$

certified by a rational function too horrendous to be reproduced here.

What has gone wrong? The problem is that while the Zeilberger algorithm and its q-analog will always find *some* recurrence satisfied by the summand that is fed in as input, there is no guarantee that the recurrence found will be the *minimal* one. And in this case, the recurrence found is *not* the minimal (second order) recurrence.

What do we do next? In order to deal with this situation, Peter Paule introduced *creative telescoping* [Pau94]. Observe that *any* function $F(k)$ can be written as a sum of its even and odd parts, i.e.,

$$F(k) = \frac{F(k) + F(-k)}{2} + \frac{F(k) - F(-k)}{2},$$

and that

$$\sum_{k=-\infty}^{\infty} \frac{F(k) - F(-k)}{2} = 0,$$

consequently,

$$f(n) := \sum_{k=\infty}^{\infty} F(n, k) = \sum_{k=\infty}^{\infty} \frac{F(n, k) + F(n, -k)}{2}.$$

Applying this observation to the $B_n^*(q)$, we find that

$$B_n^*(q) = \frac{1}{2} \sum_{k=-\infty}^{\infty} (-1)^k q^{k(5k+1)/2}(1 + q^{-k}) \binom{2n}{n + 2k}_q,$$

and upon inserting *this* presentation of the summand into the q-Zeilberger algorithm we find it does, in fact, satisfy the second order recurrence (4.35).

Thus, Paule's creative symmetrization came to the rescue in this case.

But there is no guarantee that creative symmetrization will *always* cause the Zeilberger algorithm to give the minimal order (or even a lower order) recurrence compared to the unsymmetrized version of the input.

So we may be interested in having another strategy when a non-minimal recurrence arises as output of the Zeilberger algorithm. Given that the $B_n(q)$ satisfies the second order recurrence (4.35), and the $B_n^*(q)$ satisfies the fifth order recurrence (4.38), we could prove (4.37), by showing that $B_n(q)$ *also* satisfies the fifth order recurrence (4.38) together with the initial conditions

$$B_n(q) = B_n^*(q)$$

for $n = 0, 1, 2, 3, 4$.

Thus we wish to find a third order shift operator $\mathfrak{A}_3(\eta) = a\eta^3 + b\eta^2 + c\eta + d$ such that the composition $\mathfrak{A}_3(\eta)\mathfrak{A}(\eta)$ equals the operator form of (4.38),

$$
\begin{aligned}
\mathfrak{A}^*(\eta) := {}& \eta^5 - \left(1 + q + q^2 + q^3 + q^4 + q^{2n+7} + q^{2n+8} + q^{2n+9}\right)\eta^4 \\
& + q\left(q^{4(n+3)} + q^{2(n+5)} + q^{2n+6} + q^{2n+7} + 2q^{2n+8} + 2q^{2n+9}\right. \\
& \left. + q^{4n+13} + q^{4n+14} + q^6 + q^5 + 2q^4 + 2q^3 + 2q^2 + q + 1\right)\eta^3 \\
& - q^3\left(q^{4(n+3)} + q^{2n+7} + q^{2n+8} + q^{2n+9} + q^{4n+11} + q^{6n+15}\right. \\
& \left. + q^6 + q^5 + 2q^4 + 2q^3 + 2q^2 + q + 1\right)\eta^2 \\
& - q^6\left(q^{5(n+2)} + q^{2n+3} + q^{2n+4} + q^{2n+5} + q^{5n+8}\right. \\
& \left. - q^4 - q^3 - q^2 - q - 1\right)\eta \\
& - q^{10}\left(q^{2n+1} - 1\right)\left(q^{2n+2} - 1\right). \quad (4.39)
\end{aligned}
$$

Now the composition $\mathfrak{A}_3(\eta)\mathfrak{A}(\eta)$ is of the form

$$
\begin{aligned}
\mathfrak{A}_3(\eta)\mathfrak{A}(\eta) = {}& a\eta^5 + \eta^4\left(b - a\left(-q^{n+2} + q^{2n+3} + q + 1\right)\right) \\
& + \eta^3\left(-aq^{n+2} + aq - b\left(-q^{n+2} + q^{2n+3} + q + 1\right) + c\right) \\
& + \eta^2\left(-bq^{n+2} + bq - c\left(-q^{n+2} + q^{2n+3} + q + 1\right) + d\right) \\
& + \eta\left(-cq^{n+2} + cq - d\left(-q^{n+2} + q^{2n+3} + q + 1\right)\right) \\
& - dq^{n+2} + dq, \quad (4.40)
\end{aligned}
$$

and we may equate coëfficients of η^j, $j = 0, 1, 2, 3, 4, 5$, of $\mathfrak{A}_3(\eta)\mathfrak{A}(\eta)$ with those of $\mathfrak{A}^*(\eta)$ and solve the resulting system of equations for a, b, c, d. But then we immediately see that there is a difficulty. We have four unknowns and six equations. So unless we are very lucky, and there are a sufficient number of dependencies among the equations in the 6×4 system, the system will have no solution. And alas, we are not sufficiently lucky in this case; the resulting system has no solution.

Next, we try to see if $B_n(q)$ and $B_n^*(q)$ are annihilated by a common *sixth* order operator i.e. is it the case that

$$\mathfrak{A}_1(\eta)\mathfrak{A}^*(\eta)B_n^*(q) = \mathfrak{A}_4(\eta)\mathfrak{A}(\eta)B_n(q) = 0$$

for some $\mathfrak{A}_4(\eta) = a\eta^4 + b\eta^3 + c\eta^2 + d\eta + e$ and $\mathfrak{A}_1(\eta) = f\eta + g$? In this case we are seeking a solution to a 7×7 system, so this initially seems promising, but it turns out that the only solution to this system is the trivial one, $a = b = c = d = e = f = g = 0$.

So we try one more time, but this time trying to find a *seventh* order recurrence that annihilates both $B_n(q)$ and $B_n^*(q)$.

We seek some $a, b, c, d, e, f, g, h, j$ such that

$$\mathfrak{A}_2(\eta)\mathfrak{A}^*(\eta)B_n^*(q) = \mathfrak{A}_5(\eta)\mathfrak{A}(\eta)B_n(q) = 0,$$

where $\mathfrak{A}_5(\eta) = a\eta^5 + b\eta^4 + c\eta^3 + d\eta^2 + e\eta + f$ and $\mathfrak{A}_2(\eta) = g\eta^2 + h\eta + j$.

So we now have

$$\mathfrak{A}_2(\eta)\mathfrak{A}^*(\eta) = g\eta^7$$
$$+ \left(-gq^{2n+7} - gq^{2n+8} - gq^{2n+9} - gq^4 - gq^3 - gq^2 - gq - g + h\right)\eta^6$$
$$+ \left(gq^{2n+7} - hq^{2n+7} + gq^{2n+8} - hq^{2n+8} + 2gq^{2n+9}\right.$$
$$-hq^{2n+9} + 2gq^{2n+10} + gq^{4n+14} + gq^{4n+15} + gq^{4(n+3)+1}$$
$$+gq^{2(n+5)+1} + gq^7 + gq^6 + 2gq^5 + 2gq^4$$
$$-hq^4 + 2gq^3 - hq^3 + gq^2 - hq^2$$
$$\left.+gq - hq - h + j\right)\eta^5$$
$$+ \left(hq^{2n+7} - jq^{2n+7} + hq^{2n+8} - jq^{2n+8} + 2hq^{2n+9}\right.$$
$$-jq^{2n+9} - gq^{2n+10} + 2hq^{2n+10} - gq^{2n+11}$$
$$-gq^{2n+12} - gq^{4n+14} + hq^{4n+14}$$
$$+hq^{4n+15} - gq^{6n+18} + hq^{4n+13} - gq^{4n+15}$$
$$+hq^{2n+11} - gq^9 - gq^8 - 2gq^7 + hq^7 - 2gq^6 + hq^6$$
$$-2gq^5 + 2hq^5 - gq^4 + 2hq^4 - jq^4 - gq^3 + 2hq^3 - jq^3 + hq^2$$
$$\left.-jq^2 + hq - jq - j\right)\eta^4$$
$$+ \left(jq^{2n+7} + jq^{2n+8} - gq^{2n+9} + 2jq^{2n+9} - gq^{2n+10} - hq^{2n+10}\right.$$
$$+2jq^{2n+10} - gq^{2n+11} - hq^{2n+11} - hq^{2n+12} - hq^{4n+14} + jq^{4n+14}$$
$$+jq^{4n+15} - gq^{5n+14} - hq^{6n+18} - gq^{5n+16} + jq^{4n+13}$$
$$-hq^{4n+15} + jq^{2n+11} + gq^{10} + gq^9 - hq^9 + gq^8 - hq^8$$
$$+gq^7 - 2hq^7 + jq^7 + gq^6 - 2hq^6 + jq^6$$
$$\left.-2hq^5 + 2jq^5 - hq^4 + 2jq^4 - hq^3 + 2jq^3 + jq^2 + jq\right)\eta^3$$
$$+ \left(-hq^{2n+9} - hq^{2n+10} - jq^{2n+10} + gq^{2n+11} - hq^{2n+11} - jq^{2n+11}\right.$$
$$+gq^{2n+12} - jq^{2n+12} - gq^{4n+13} - jq^{4n+14}$$
$$-hq^{5n+14} - jq^{6n+18} - hq^{5n+16} - jq^{4n+15}$$
$$-gq^{10} + hq^{10} + hq^9 - jq^9 + hq^8 - jq^8 + hq^7 - 2jq^7 + hq^6 - 2jq^6$$

$$-2jq^5 - jq^4 - jq^3\big)\,\eta^2$$
$$+ \left(-jq^{2n+9} - jq^{2n+10} + hq^{2n+11} - jq^{2n+11} + hq^{2n+12} - hq^{4n+13}\right.$$
$$\left.-jq^{5n+14} - jq^{5(n+2)+6} - hq^{10} + jq^{10} + jq^9 + jq^8 + jq^7 + jq^6\right)\eta$$
$$+ jq^{2n+11} + jq^{2n+12} - jq^{4n+13} - jq^{10} \quad (4.41)$$

and

$$\mathfrak{A}_5(\eta)\mathfrak{A}(\eta) = \eta^5\left(-aq^{n+2} + aq - b\left(-q^{n+2} + q^{2n+3} + q + 1\right) + c\right)$$
$$+ \eta^6\left(b - a\left(-q^{n+2} + q^{2n+3} + q + 1\right)\right) + a\eta^7$$
$$+ \eta^4\left(-bq^{n+2} + bq - c\left(-q^{n+2} + q^{2n+3} + q + 1\right) + d\right)$$
$$+ \eta^3\left(-cq^{n+2} + cq - d\left(-q^{n+2} + q^{2n+3} + q + 1\right) + e\right)$$
$$+ \eta^2\left(-dq^{n+2} + dq - e\left(-q^{n+2} + q^{2n+3} + q + 1\right) + f\right)$$
$$+ \eta\left(-eq^{n+2} + eq - f\left(-q^{n+2} + q^{2n+3} + q + 1\right)\right)$$
$$- fq^{n+2} + fq. \quad (4.42)$$

Equating coëfficients of η^r for $r = 0, 1, \ldots, 7$, and solving the resulting system of 8 equations in 9 unknowns, we at last obtain a nontrivial set of solutions. We may assign the following values to $a, b, c, \ldots$, in the above expressions to obtain a seventh order annihilating operator. In what follows, t is a free parameter.

$$a = t$$
$$b = -t\left(q^{2n+7} + q^{2n+8} + q^{2n+9} + q^4 + q^3 + q^2 + q + 1\right)$$
$$c = qt\left(q^{2n+6} + q^{2n+7} + 2q^{2n+8} + 2q^{2n+9} + q^{2n+10} + q^{4n+12}\right.$$
$$\left.+q^{4n+13} + q^{4n+14} + q^6 + q^5 + 2q^4 + 2q^3 + 2q^2 + q + 1\right)$$
$$d = -q^3 t\left(q^{2n+7} + q^{2n+8} + q^{2n+9} + q^{4n+11} + q^{4n+12} + q^{6n+15}\right.$$
$$\left.+q^6 + q^5 + 2q^4 + 2q^3 + 2q^2 + q + 1\right)$$
$$e = -tq^6\left(q^{2n+3} + q^{2n+4} + q^{2n+5} + q^{5n+8} + q^{5n+10}\right.$$
$$\left.-q^4 - q^3 - q^2 - q - 1\right)$$
$$f = -q^{10}t\left(-q^{2n+1} - q^{2n+2} + q^{4n+3} + 1\right)$$
$$g = t$$
$$h = -t(1 + q - q^{n+2} + q^{2n+3})$$
$$j = tq(1 - q^{n+1}).$$

And thus, upon testing the initial conditions $B_n(q) = B_n^*(q)$ for $0 \leqq n \leqq 6$, Bressoud's identity is once again established, and perseverance has reaped its reward.

Bressoud's polynomial generalization of the second Rogers–Ramanujan identity [Bre81, p. 212, Eq. (1.5)] may be stated as

$$\sum_{k=0}^{n} q^{k(k+1)} \binom{n}{k}_q = \sum_{k \in \mathbb{Z}} (-1)^k q^{k(5j+3)/2} \binom{2n+1}{n+2k+1}_q. \tag{4.43}$$

Eq. (4.43) can be proved similarly to Bressoud's polynomial generalization of the first Rogers–Ramanujan identity.

4.4   Finite Ramanujan–Slater Mod 8 identities

We now apply the method of q-difference equations to find a finite version of the first Ramanujan–Slater identity (related to the first Göllnitz–Gordon partition identity). Recall that Eq. (2.16) is

$$R(q) := \sum_{j=0}^{\infty} \frac{q^{j^2}(-q;q^2)_j}{(q^2;q^2)_j} = \frac{f(-q^3,-q^5)}{\psi(-q)}.$$

As usual, we begin with the series side of the identity, and insert the second variable t in such a way to allow us to obtain a nonhomogeneous q-difference equation. We define

$$\mathfrak{R}(t) := \mathfrak{R}(t,q) := \sum_{j=0}^{\infty} \frac{t^j q^{j^2}(-tq;q^2)_j}{(t;q^2)_{j+1}}. \tag{4.44}$$

Remark 4.2. The alert reader may at this point ask why we did not instead define $\mathfrak{R}(t)$ to be the series

$$\sum_{j=0}^{\infty} \frac{t^{2j} q^{j^2}(-t^2q;q^2)_j}{(t^2;q^2)_{j+1}}, \tag{4.45}$$

which would appear to be more consistent with how we formed $\mathfrak{G}(t)$ from $G(q)$. The reason is that when we expand the series (4.45) as a power series in t (with coëfficients that are polynomials in q), only even powers of t will appear. While there is nothing "wrong" with that, it seems nicer to let all integer powers of t appear; so we replace t by $t^{1/2}$ in Eq. (4.45) to obtain (4.44).

Next, we split off the $j = 0$ term and form the corresponding q-difference equation:

$$\mathfrak{R}(t) = \frac{1}{1-t} + \sum_{j=1}^{\infty} \frac{t^j q^{j^2}(-tq;q^2)_j}{(t;q^2)_{j+1}}$$

$$= \frac{1}{1-t} + \sum_{j=0}^{\infty} \frac{t^{j+1} q^{j^2+2j+1}(-tq;q^2)_{j+1}}{(t;q^2)_{j+2}}$$

$$= \frac{1}{1-t} + \frac{tq}{1-t} \sum_{j=0}^{\infty} \frac{(tq^2)^j q^{j^2}(-(tq^2)q;q^2)_j}{(tq^2;q^2)_{j+1}}$$

$$= \frac{1}{1-t} + \frac{tq}{1-t} \mathfrak{R}(tq^2).$$

Thus

$$(1-t)\mathfrak{R}(t) = 1 + tq\mathfrak{R}(tq^2),$$

or

$$\mathfrak{R}(t) = 1 + t\mathfrak{R}(t) + tq\mathfrak{R}(tq^2).$$

Thinking of $\mathfrak{R}(t) = \sum_{n=0}^{\infty} P_n(q)t^n$, we deduce

$$P_0(q) = 1, \qquad P_1(q) = 1 + q \tag{4.46}$$

and

$$P_n(q) = (1 + q^{2n-1})P_{n-1}(q) + q^{2n-2}P_{n-2}(q), \tag{4.47}$$

if $n \geq 2$. (We note in passing that the $P_n(q)$ are q-analogs of the Pell numbers.) Also

$$\sum_{n=0}^{\infty} P_n(q)t^n = \mathfrak{R}(t) = \sum_{j=0}^{\infty} \frac{t^j q^{j^2}(-tq;q^2)_j}{(t;q^2)_{j+1}}$$

$$= \sum_{j=0}^{\infty} t^j q^{j^2} \sum_{k=0}^{j} \binom{j}{k}_{q^2} (-1)^k (-tq)^k q^{k(k-1)} \sum_{l=0}^{\infty} \binom{l+j}{j}_{q^2} t^l$$

$$\text{(by (1.56) and (1.80))}$$

$$= \sum_{j,k,l\geq 0} t^{j+k+l} q^{k^2+j^2} \binom{j}{k}_{q^2} \binom{j+l}{j}_{q^2}$$

$$= \sum_{n\geq 0} t^n \sum_{j,k\geq 0} q^{j^2+k^2} \binom{j}{k}_{q^2} \binom{n-k}{j}_{q^2}.$$

Thus we have

$$P_n(q) = \sum_{j,k\geq 0} q^{j^2+k^2} \binom{j}{k}_{q^2} \binom{n-k}{j}_{q^2},$$

and

$$\lim_{n\to\infty} P_n(q) = R(q).$$

As before, we seek a second representation of $P_n(q)$ for which it is easy to see that

$$\lim_{n\to\infty} P_n(q) = \frac{f(-q^3, -q^5)}{\psi(-q)}.$$

Unlike our previous example, it is not the q-binomial coëfficient that plays a central role here, but rather a sum of two adjacent q-trinomial coëfficients. We define

$$U(n, j; q) := T_0(n, j; q) + T_0(n, j+1; q),$$

where $T_0(n, j; q)$ is defined in Eq. (1.91). It can be shown ([AB87, §2.3] and [And90a]) that

$$\lim_{n \to \infty} U(n, j; q) = \frac{1}{\psi(-q)}.$$

With some effort, it is reasonable to conjecture that

$$P_n(q) = \sum_{j \in \mathbb{Z}} (-1)^j q^{4j^2 + j} U(n, 4j; q).$$

And thus

$$\sum_{j,k \geq 0} q^{j^2 + k^2} \binom{j}{k}_{q^2} \binom{n-k}{j}_{q^2} = \sum_{j \in \mathbb{Z}} (-1)^j q^{4j^2 + j} U(n, 4j; q)$$

may be established as a finite form of (2.16), upon showing that the right-hand side satisfies the recurrence (4.47) and initial conditions (4.46).

The details are provided by Andrews [And90b, §4, pp. 9–14]. The guiding philosophy of the proof is the same as we encountered with the finite Rogers–Ramanujan identities. The only technical difference is that rather than using the q-Pascal triangle type recurrences for the q-binomial coëfficients, one must use the analogous recurrences for the T_0 and T_1 q-trinomial coëfficients (see [AB87]):

$$T_1(n, j; q) = T_1(n-1, j; q) + q^{n+j} T_0(n-1, j+1; q) + q^{n-j} T_0(n-1, j-1; q), \tag{4.48}$$

$$T_0(n, j; q) = T_0(n-1, j-1; q) + q^{n+j} T_1(n-1, j; q) + q^{2n+2j} T_0(n-1, j+1; q), \tag{4.49}$$

$$T_1(n, j; q) - q^{n-j} T_0(n, j; q) - T_1(n, j+1; q) + q^{n+j+1} T_0(n, j+1; q) = 0. \tag{4.50}$$

4.5   Borwein Conjecture

In 1990, Peter Borwein posed the following conjecture to George Andrews in a phone call.

Conjecture 4.3. *Define polynomials $A_n(q)$, $B_n(q)$, $C_n(q)$ so that*

$$(q;q^3)_n(q^2;q^3)_n = A_n(q^3) - qB_n(q^3) - q^2C_n(q^3).$$

Then all of the coëfficients of $A_n(q)$, $B_n(q)$, $C_n(q)$ are nonnegative.

One can show that these polynomials have the following representations

$$A_n(q) = \sum_{j=-\infty}^{\infty} (-1)^j q^{j(9j+1)/2}\binom{2n}{n+3j}_q,\qquad(4.51)$$

$$B_n(q) = \sum_{j=-\infty}^{\infty} (-1)^j q^{j(9j-5)/2}\binom{2n}{n+3j-1}_q,\qquad(4.52)$$

$$C_n(q) = \sum_{j=-\infty}^{\infty} (-1)^j q^{j(9j+7)/2}\binom{2n}{n+3j+1}_q,\qquad(4.53)$$

and

$$A_n(q) = \sum_{0\le 3j\le n} \frac{(q^3;q^3)_{n-j-1}(1-q^{2n})(q;q)_{3j}q^{3j^2}}{(q;q)_{n-3j}(q^3;q^3)_{2j}(q^3;q^3)_j},\qquad(4.54)$$

$$B_n(q) = \sum_{0\le 3j\le n-1} \frac{(q^3;q^3)_{n-j-1}(1-q^{3j+2}-q^{n+3j+2}+q^{n+1})(q;q)_{3j}q^{3j^2+3j}}{(q;q)_{n-3j-1}(q^3;q^3)_{2j+1}(q^3;q^3)_j},$$
$$(4.55)$$

$$C_n(q) = \sum_{0\le 3j\le n-1} \frac{(q^3;q^3)_{n-j-1}(1-q^{3j+1}-q^{n+3j+2}+q^{n})(q;q)_{3j}q^{3j^2+3j}}{(q;q)_{n-3j-1}(q^3;q^3)_{2j+1}(q^3;q^3)_j}.$$
$$(4.56)$$

Nonetheless, knowledge of these representations does not seem to lead to a proof of the conjecture.

4.6   Exercises

1. Prove the recurrences (4.48)–(4.50), directly from the definitions of $T_0(n,j;q)$ and $T_1(n,j;q)$.

2. Better yet, find automated proofs of the recurrences (4.48)–(4.50) using your favorite implementation of the q-Zeilberger algorithm!

3. Prove that the polynomials $A_n(q)$, $B_n(q)$, $C_n(q)$ satisfy the following recurrences for $n \ge 1$.

(a) $A_n(q) = (1 + q^{2n-1})A_{n-1}(q) + q^n B_{n-1}(q) + q^n C_{n-1}(q),$

(b) $B_n(q) = (1 + q^{2n-1})B_{n-1}(q) + q^{n-1}A_{n-1}(q) - q^n C_{n-1}(q),$

(c) $C_n(q) = (1 + q^{2n-1})C_{n-1}(q) + q^{n-1}A_{n-1}(q) - q^{n-1}B_{n-1}(q),$

(d) $A_0(q) = 1,\ B_0(q) = C_0(q) = 0.$

4. Prove Eqs. (4.51)–(4.56). Again, the q-Zeilberger algorithm is highly recommended.

5. Show that letting $n \to \infty$ in (4.51)–(4.56) yields the Bailey identities related to the modulus 9 (2.18)–(2.20).

Chapter 5

Motivated Proofs, Connections to Lie Algebras, and More Identities

5.1 "Motivated" Proofs

5.1.1 Rogers–Ramanujan Products Without the Identities

As we have seen, the classic proofs of the Rogers–Ramanujan identities are, as Hardy pointed out, "essentially verifications." We now wish to reëxamine the Rogers–Ramanujan identities from a somewhat different perspective.

Let

$$G_1(q) := \frac{1}{(q;q^5)_\infty (q^4;q^5)_\infty}$$

and

$$G_2(q) := \frac{1}{(q^2;q^5)_\infty (q^3;q^5)_\infty}.$$

Next, let

$$G_3(q) := \frac{G_1(q) - G_2(q)}{q}.$$

Now consider: *if we know the combinatorial version of the Rogers–Ramanujan identities*, it trivially follows that $G_3(q)$ is a power series with nonnegative coëfficients. Why? Because the identities tell us that $G_1(q)$ is the generating function for $r_1(n)$, the number of partitions of n into parts that mutually differ by at least 2, and that $G_2(q)$ is the generating function for $r_2(n)$, the number of partitions of n into parts which mutually differ by at least 2 and are each greater than 1. Thus $G_3(n)$ is q^{-1} times the generating function for $r_1(n) - r_2(n)$, and since, for each positive integer n, those partitions enumerated by $r_2(n)$ are clearly a subset of those partitions enumerated by $r_1(n)$, it must be that $r_1(n) - r_2(n) \geqq 0$. Further, since $r_1(0) = r_2(0) = 1$, the constant term of $G_1(q) - G_2(q)$ is 0, so no negative power of q is introduced when $G_1(q) - G_2(q)$ is divided by q.

Likewise, *if we know the series-product version of the Rogers–Ramanujan identities*, we know that

$$G_1(q) = \sum_{j=0}^{\infty} \frac{q^{j^2}}{(q;q)_j},$$

139

and

$$G_2(q) = \sum_{j=0}^{\infty} \frac{q^{j^2+j}}{(q;q)_j},$$

and thus

$$G_3(q) = \frac{G_1(q) - G_2(q)}{q}$$

$$= q^{-1} \sum_{j=0}^{\infty} \left(\frac{q^{j^2}}{(q;q)_j} - \frac{q^{j^2+j}}{(q;q)_j} \right)$$

$$= q^{-1} \sum_{j=0}^{\infty} \frac{q^{j^2}(1-q^j)}{(q;q)_j}$$

$$= \sum_{j=0}^{\infty} \frac{q^{j^2-1}}{(q;q)_{j-1}}$$

$$= \sum_{j=0}^{\infty} \frac{q^{j^2+2j}}{(q;q)_j},$$

which again clearly has no negative coëfficients of q^n nor negative powers of q.

However, if we do not know the Rogers–Ramanujan identities, then it is not at all obvious that

$$G_3(q) = \frac{G_1(q) - G_2(q)}{q}$$

$$= q^{-1} \left(\frac{1}{(q;q^5)_\infty (q^4;q^5)_\infty} - \frac{1}{(q^2;q^5)_\infty (q^3;q^5)_\infty} \right) \quad (5.1)$$

has a power series expansion with nonnegative coëfficients.

In fact, at the 1987 AMS Institute on Theta Functions, Leon Ehrenpreis of Temple University asked whether it was possible to prove that (5.1) has a power series expansion with nonnegative coëfficients, without employing the Rogers–Ramanujan identities [AB89, p. 402].

Andrews and Baxter answered Ehrenpreis' question in the affirmative in their 1989 paper in the *Monthly* [AB89]. We shall follow their development closely here.

5.1.2 Ehrenpreis' Question and the Empirical Hypothesis

We note that

$$G_1(q) = 1 + q + q^2 + q^3 + 2q^4 + 2q^5 + 3q^6 + 3q^7 + 4q^8 + 5q^9$$
$$+ 6q^{10} + 7q^{11} + 9q^{12} + \cdots,$$

$$G_2(q) = 1 + q^2 + q^3 + q^4 + q^5 + 2q^6 + 2q^7 + 3q^8 + 3q^9$$
$$+ 4q^{10} + 4q^{11} + 6q^{12} + \cdots,$$
$$G_3(q) = 1 + q^3 + q^4 + q^5 + q^6 + q^7 + 2q^8 + 2q^9$$
$$+ 3q^{10} + 3q^{11} + 4q^{12} + \cdots.$$

If we next subtract $G_3(q)$ from $G_2(q)$, we obtain a series with lowest degree term q^2. Let us therefore define

$$G_4(q) := (G_2(q) - G_3(q))/q^2.$$

Then

$$G_4(q) = 1 + q^4 + q^5 + q^6 + q^7 + q^8 + q^9 + 2q^{10} + 2q^{11}$$
$$+ 3q^{12} + 3q^{13} + 4q^{14} + \cdots.$$

A similar observation motivates us to define

$$G_5(q) := (G_3(q) - G_4(q))/q^3,$$

and more generally

$$G_i(q) := \frac{G_{i-2}(q) - G_{i-1}(q)}{q^{i-2}}, \qquad (5.2)$$

for $i \geq 3$.

We then allow our computer algebra system to tell us that

$$G_5(q) = 1 + q^5 + q^6 + q^7 + q^8 + q^9 + q^{10} + q^{11} + 2q^{12} + \cdots$$
$$G_6(q) = 1 + q^6 + q^7 + q^8 + q^9 + q^{10} + q^{11} + q^{12} + q^{13} + \cdots$$
$$G_7(q) = 1 + q^7 + q^8 + q^9 + q^{10} + q^{11} + q^{12} + q^{13} + q^{14} + \cdots$$
$$G_8(q) = 1 + q^8 + q^9 + q^{10} + q^{11} + q^{12} + q^{13} + q^{14} + q^{15} + \cdots$$

$$\vdots$$

Based on the above we make the following conjecture that Andrews and Baxter called their "Empirical Hypothesis." We designate it a proposition here, as it will be proved momentarily.

Proposition 5.1 (The AB Empirical Hypothesis). *For $i \geq 1$,*

$$G_i(q) = 1 + \sum_{n=i}^{\infty} g_{i,n} q^n \qquad (5.3)$$

for some positive integers $g_{i,n}$.

Remember, our immediate goal is to answer Ehrenpreis' question in the affirmative i.e. we want to show that the power series expansion of $G_3(q)$ has all nonnegative coëfficients. To this end, we wish to write $G_3(q)$ as a linear combination of $G_i(q)$ and $G_{i+1}(q)$ for all $i \geq 3$. So next we calculate the required linear combination. We can rearrange (5.2) as

$$G_{i-2}(q) = G_{i-1} + q^{i-2}G_i(q), \tag{5.4}$$

where $i \geq 3$.

By repeated use of (5.2) with larger and larger values of i, we obtain

$$
\begin{aligned}
G_3(q) &= G_4(q) + q^3 G_5(q) \\
&= G_5(q) + q^4 G_6(q) + q^3 G_5(q) \\
&= (1 + q^3)G_5(q) + q^4 G_6(q) \\
&= (1 + q^3 + q^4)G_6(q) + q^5(1 + q^3)G_7(q).
\end{aligned}
$$

Continue the above process to learn that

$$G_3(q) = A_i(q)G_i(q) + B_i(q)G_{i+1}(q), \tag{5.5}$$

where, by Eq. (5.4),
$$A_{i+1}(q) = A_i(q) + B_i(q) \tag{5.6}$$

and
$$B_{i+1}(q) = q^i A_i(q) \tag{5.7}$$

with $A_3(q) = 1$, $B_3(q) = 0$.

But we may eliminate $B_i(q)$ from the preceding two equations to find

$$A_{i+1}(q) = A_i(q) + q^{i-1}A_{i-1}(q), \tag{5.8}$$

when $i \geq 4$, and $A_3(q) = A_4(q) = 1$.

Now we have in hand the recurrence and initial conditions necessary to see that $\lim_{i\to\infty} A_i(q)$ is a power series with nonnegative coëfficients. Thus, provided that the AB empirical hypothesis (5.3) holds, we may use (5.5) to deduce

$$
\begin{aligned}
G_3(q) &= \lim_{i\to\infty} \left(A_i(q)G_i(q) + q^{i-1}A_{i-1}(q)G_{i+1}(q) \right) \\
&= \left(\lim_{i\to\infty} A_i(q) \right) \left(\lim_{i\to\infty} G_i(q) \right) \\
&\quad + 0 \left(\lim_{i\to\infty} A_{i-1}(q) \right) \left(\lim_{i\to\infty} G_{i+1}(q) \right) \\
&= A_\infty(q)
\end{aligned}
$$

Thus, all that remains to dispense with Ehrenpreis' question is the proof of Proposition 5.1.

Proof of Proposition 5.1. We may use the Jacobi triple product identity (1.58) to find

$$G_1(q) = \frac{f(-q^2, -q^3)}{(q;q)_\infty} = \frac{\sum_{j=-\infty}^{\infty}(-1)^j q^{j(5j-1)/2}}{(q;q)_\infty}$$

$$= \frac{1 + \sum_{j=1}^{\infty}(-1)^j q^{j(5j-1)/2}(1+q^j)}{(q;q)_\infty} \quad (5.9)$$

and

$$G_2(q) = \frac{f(-q, -q^4)}{(q;q)_\infty} = \frac{\sum_{j=-\infty}^{\infty}(-1)^j q^{j(5j-3)/2}}{(q;q)_\infty}$$

$$= \frac{\sum_{j=0}^{\infty}(-1)^j q^{j(5j+3)/2}(1-q^{2j-1})}{(q;q)_\infty}. \quad (5.10)$$

Using these representations of $G_1(q)$ and $G_2(q)$, we may (with some effort) deduce that

$$qG_3(q) = \frac{\sum_{j=0}^{\infty}(1-q^{j+1})(1-q^{2j+2})(-1)^j q^{6j+5j(j-1)/2}}{(q;q)_\infty}. \quad (5.11)$$

It should be noted that in order to *see* the desired form (5.11), it is helpful to expand the numerators of (5.9) and (5.10) using a computer algebra system, perform the subtraction, and then group terms appropriately.

Similarly (again with the aid of your favorite computer algebra software), we may find that

$$q^2 G_4(q) = G_2(q) - G_3(q)$$

$$= \frac{1}{(q;q)_\infty}\sum_{j=0}^{\infty}(1-q^{j+1})(1-q^{j+2})(1-q^{2j+3})(-1)^j q^{8j+5j(j-1)/2}. \quad (5.12)$$

After several more such iterations, it is natural to conjecture that

$$q^{i-2}G_i(q) = \frac{1}{(q;q)_\infty}\sum_{j=0}^{\infty}(q^{j+1};q)_{i-2}(-1)^j q^{2ij+5j(j-1)/2}(1-q^{2j+i-1}). \quad (5.13)$$

To establish (5.13), we may proceed by induction. We have already verified the base cases. So suppose (5.13) holds for i and $i+1$. Then

$$(q;q)_\infty (G_i(q) - G_{i+1}(q))$$

$$= (q;q)_{i-1} + \sum_{j=1}^{\infty}(q^{j+1};q)_{i-2}(-1)^j q^{2ij+5j(j-1)/2}(1-q^{2j+i-1})$$

$$- \left\{ \sum_{j=0}^{\infty} (q^{j+1}; q)_{i-1} (-1)^j q^{2(i+1)j+5j(j-1)/2} \right.$$

$$\left. - \sum_{j=0}^{\infty} (q^{j+1}; q)_{i-1} (-1)^j q^{2(i+1)j+5j(j-1)/2+2j+i} \right\}$$

$$= - \sum_{j=0}^{\infty} (q^{j+2}; q)_{i-2} (-1)^j q^{2i(j+1)+5j(j+1)/2} (1 - q^{2j+i+1})$$

$$+ \sum_{j=0}^{\infty} (q^{j+2}; q)_{i-1} (-1)^j q^{2(i+1)(j+1)+5j(j+1)/2}$$

$$+ \sum_{j=0}^{\infty} (q^{j+1}; q)_{i-1} (-1)^j q^{2(i+1)j+5j(j-1)/2+2j+i}$$

$$= q^i \sum_{j=0}^{\infty} (q^{j+2}; q)_{i-2} (-1)^j q^{2(i+1)j+5j(j-1)/2+2j}$$

$$\{-q^{i+j}(1 - q^{2j+i+i}) + q^{i+3j+2}(1 - q^{i+j}) + (1 - q^{j+1})\}$$

$$= q^i \sum_{j=0}^{\infty} (q^{j+1}; q)_i (-1)^j q^{2(i+1)j+5j(j-1)/2}$$

$$= q^i (q; q)_\infty G_{i+2}(q).$$

Thus (5.13) is valid for $i \geqq 2$, and so it follows that

$$G_i(q)$$

$$= \frac{1}{(q^i; q)_\infty} + \frac{1}{(q; q)_\infty} \sum_{j=1}^{\infty} (q^{j+1}; q)_{i-2} (-1)^j q^{2ij+5j(j-1)/2} (1 - q^{2j+i-1})$$

$$= 1 + q^i \gamma_i(q),$$

for some power series $\gamma_i(q)$. $\square$

5.1.3   Rogers–Ramanujan Identities

Having dispensed with Ehrenpreis' question, we now turn to the question of how one might start with the infinite products $G_1(q)$ and $G_2(q)$ and be led to the Rogers–Ramanujan identities.

If we start with $G_1(q)$ instead of $G_3(q)$ as our main focus, we are led down essentially the same path including the fact that an analogous $A_i(q)$ satisfies the same recurrence (5.8); only the initial conditions differ. In this case, we have initial conditions $A_2(q) = 1$, $A_3(q) = 1+q$. By iterating (5.8), we obtain

$$A_4(q) = 1 + q^1 + q^2$$

$$A_5(q) = 1 + q^1 + q^2 + q^3 + q^{3+1}$$
$$A_6(q) = 1 + q^1 + q^2 + q^3 + q^{3+1} + q^4 + q^{4+1} + q^{4+2}$$

$$\vdots$$

And thus by starting out with just the product representations for G_1 and G_2, the resulting recurrence reveals the hint of the partitions with difference at least 2 in the exponents of the resulting sequence of polynomials $A_n(q)$. Further, these examples suggest that $A_{i+1}(q)$ is the generating function for partitions into parts less than i where all parts must differ from each other by at least 2. This can be proved by induction, as the first term on the right-hand side of (5.8), $A_i(q)$, generates the difference-2 partitions with parts $<$ $i - 1$, while the second term, $q^{i-1}A_{i-1}(q)$ generates partitions with largest part exactly equal to $i - 1$ and all parts differing by at least 2.

Then we note that $G_1(q) = A_\infty(q)$, which is the first Rogers–Ramanujan identity. The second Rogers–Ramanujan identity follows analogously.

5.1.4   Generalizations

In [LZ12], J. Lepowsky and M. Zhu extended the Andrews–Baxter "motivated" proof of the Rogers–Ramaunjan identities to give an analogously "motivated" proof of the Gordon–Andrews generalization of the Rogers–Ramanujan identities to all odd moduli. Later, Kanade, Lepowsky, Russell, and Sills [KLRS17] found the analogous "motivated" proof of the Bressoud's even modulus analog of Gordon–Andrews. The surprise here was the necessity of introducing "ghost series," in which the associated partitions satisfied the ordinary difference conditions, but satisfied the opposite of the parity condition from Bressoud's identity.

Additionally, a group of seven collaborating at Rutgers (Coulson, Kanade, Lepowsky, McRae, Qi, Russell, and Sadowski) published [CKL+17] an analogous "motivated" proof of the Andrews infinite family generalization of the Göllnitz–Gordon identities, Theorem 3.13.

These "motivated" proofs are of central importance to Lepowsky and others working in Lie and vertex operator algebra theory. We explore why in the next section.

5.2   Connections to Lie and Vertex Operator Algebras

5.2.1   The 1970s Through the Mid-1990s

In the late 1970s, Lepowsky (as an assistant professor) and Steve Milne (as a postdoc) at Yale , were interested in the representation theory of Lie algebras and became aware of George Andrews' new (at the time) book *The Theory of Partitions* [And76]. The Lepowsky–Milne collaboration led to [LM78]. Further, Robert Wilson spent his first sabbatical year from Rutgers, 1976–1977, at Yale, and just afterwards, Lepowsky joined the faculty at Rutgers allowing the Lepowsky–Wilson collaboration to blossom.

Lepowsky and Wilson gave the first two Lie theoretic (or, more preceisely, vertex operator theoretic) proofs of the Rogers–Ramanujan identities [LW81, LW82, LW84], using their theory of twisted vertex operators which they developed for this purpose starting with [LW78].

Here they showed that the sum sides (i.e. the set of partitions where parts must differ by at least two) of the two Rogers–Ramanujan identities corresponded in a precise way to the two inequivalent level 3 standard modules of the affine Kac–Moody Lie algebra $A_1^{(1)}$. Lepowsky and Wilson went on to show that the standard modules of $A_1^{(1)}$ of odd level correspond with the Andrews–Gordon generalization of the Rogers–Ramanujan identities to all odd moduli (Eq. (3.12)) and the even level standard modules of $A_1^{(1)}$ correspond to Bressoud's even modulus analog of Andrews–Gordon (Eq. (3.13)) [LW84, LW85].

We shall not attempt to explain this correspondence between the standard modules of $A_1^{(1)}$ and the Andrews–Gordon–Bressoud identities in any real detail here. Suffice it to state that there is a quantity called the *principal character* of a standard module of an affine Kac–Moody Lie algebra that turns out to be equal to an infinite product that may correspond to the product side of a Rogers–Ramanujan type identity. The intrigued reader is encouraged to consult Lepowsky's article [Lep07], along with the previously cited references for details.

Once $A_1^{(1)}$ was "understood" in terms of Andrews–Gordon–Bressoud, the next obvious project to try was an analogous analysis of the standard modules of $A_2^{(2)}$. This project was assigned to Lepowsky and Wilson's PhD student Stefano Capparelli. It turned out that $A_2^{(2)}$ was *much* more difficult than $A_1^{(1)}$. As it happens, the level 2 standard modules of $A_2^{(2)}$ are related to the Rogers–Ramanujan identities with q replaced by q^2 throughout. The two inequivalent level 3 standard modules, after much effort by Capparelli, yielded a pair of brand new partition identities in the spirit of Rogers–Ramanujan and Schur, but more complicated. They are as follows.

Theorem 5.2 (Capparelli's first partition identity). *Let $C_1(n)$ denote the number of partitions $\lambda = (\lambda_1, \lambda_2, \ldots, \lambda_l)$ (l arbitrary) of n wherein*

- $\lambda_i - \lambda_{i+1} \geqq 2$,

- $\lambda_i - \lambda_{i+1} = 2$ *only if* $\lambda_i \equiv 1 \pmod 3$,

- $\lambda_i - \lambda_{i+1} = 3$ *only if* $\lambda_i \equiv 0 \pmod 3$,

- $\lambda_i \neq 1$ *for any i .*

Let $C_2(n)$ denote the number of partitions of n into parts $\equiv \pm 2, \pm 3 \pmod{12}$. Let $C_3(n)$ denote the number of partitions of n into distinct parts $\not\equiv \pm 1 \pmod 6$. Then $C_1(n) = C_2(n) = C_3(n)$ for all n.

Theorem 5.3 (Capparelli's second partition identity). *Let $D_1(n)$ denote the number of partitions $\lambda = (\lambda_1, \lambda_2, \ldots, \lambda_l)$ (l arbitrary) of n wherein*

- $\lambda_i - \lambda_{i+1} \geq 2$,

- $\lambda_i - \lambda_{i+1} = 2$ only if $\lambda_i \equiv 1 \pmod 3$,

- $\lambda_i - \lambda_{i+1} = 3$ only if $\lambda_i \equiv 0 \pmod 3$,

- $\lambda_i \neq 2$ for any i .

Let $D_2(n)$ denote the number of partitions of n into distinct parts $\not\equiv \pm 2 \pmod 6$. Then $D_1(n) = D_2(n)$ for all n.

Capparelli earned his PhD in 1988 with a thesis that contained Theorems 5.2 and 5.3 still as conjectures. Lepowsky announced the first of Capparelli's conjectured identities in his talk at the Centenary Conference in honor of Hans Rademacher, held at Penn State from July 21–25, 1992. As it happened, George Andrews, Basil Gordon, and David Bressoud were all in the audience for Lepowsky's talk and were very intrigued by these new conjectural identities. Andrews spent the next several evenings working to find a proof of the identities, and managed to do so just in time to present it as his talk on the last day of the conference. This proof is presented in [And94]. Later, vertex operator theoretic proofs were given by M. Tamba and C. Xie [TX95] and independently by Capparelli [Cap96].

The simplest proof is the one due to Andrews and relies on recurrences, and is similar in spirit to the proof of Theorem 1.66.

Remark 5.4. The difference conditions on the parts that characterize the partitions enumerated by $C_1(n)$ and $D_1(n)$ are more complicated, but in the same spirit as Schur's mod 6 partition theorem. It is interesting to note that the statement of the difference conditions can be simplified if we *reorder* the positive integers as follows:

$$2 < 1 < 3 < 4 < 5 < 6$$
$$< 8 < 7 < 9 < 10 < 11 < 12$$
$$< 14 < 13 < 15 < 16 < 17 < 18 < \ldots,$$

i.e. transpose the positions of adjacent integers that are congruent to 1 and 2 (mod 6), and leave the positions of all other integers unchanged. Then a single difference condition characterizes the partitions enumerated by $C_1(n)$ and $D_1(n)$, namely that λ_i and λ_{i+1} must be separated by at least three positions.

We note that Alladi and Gordon introduced the *method of weighted words* in [AG93] and used this method to great effect in finding refinements and generalizations of the Schur and Capparelli theorems [AG95, AAG95] (the latter paper is joint with Andrews). Different orderings of the integers are seen to give rise to different companion partition identities. However, in [AAG95, pp. 647–648], we find that it is the *standard* ordering of the integers $1 < 2 < 3 < 4 < 5 < \dots$ that gives rise to the *original* formulation of Capparelli's identity (as stated in Theorem 5.2), and thus the re-ordering of the positive integers suggested here is different.

5.2.2   Conjectured Partition Identities

Many years would pass before the analogous partition identities for level 4 would be discovered. The breakthrough for level 4 is due to Debajyoti Nandi [Nan14], the last PhD student of Wilson before he retired in May 2014, after completing forty-three years on the faculty of Rutgers. There are three inequivalent level 4 standard modules of $A_2^{(2)}$; each corresponds to a partition identity. The conjectured identity associated with the $(4,0)$ module is as follows.

Conjecture 5.5 (Nandi). *The number of partitions of n into parts $\equiv \pm 2, \pm 3, \pm 4 \pmod{14}$ equals the number of partitions $\lambda = (\lambda_1, \lambda_2, \dots, \lambda_l)$ of n where*

- $m_1(\lambda) = 0$,

- $\lambda_i - \lambda_{i+1} \neq 1$,

- $\lambda_i - \lambda_{i+2} \geqq 3$,

- $\lambda_i - \lambda_{i+2} = 3 \implies \lambda_i \neq \lambda_{i+1}$,

- $\lambda_i - \lambda_{i+2} = 3$ *and* $2 \nmid \lambda_i \implies \lambda_{i+1} \neq \lambda_{i+2}$.

- $\lambda_i - \lambda_{i+2} = 4$ *and* $2 \nmid \lambda_i \implies \lambda_i \neq \lambda_{i+1}$,

- *Consider the first differences*

$$\Delta \lambda := (\lambda_1 - \lambda_2, \lambda_2 - \lambda_3, \dots, \lambda_{l-1} - \lambda_l).$$

None of the following subwords are permitted in $\Delta \lambda$:

$$(3,3,0), (3,2,3,0), (3,2,2,3,0), \dots, (3,2,2,2,2,\dots,2,3,0).$$

The conjectured identity associated with the $(2,1)$-module is given next.

Conjecture 5.6 (Nandi). *The number of partitions of n into parts $\equiv \pm 1, \pm 4, \pm 6 \pmod{14}$ equals the number of partitions $\lambda = (\lambda_1, \lambda_2, \dots, \lambda_l)$ of n where*

- $m_2(\lambda) = 0$,

- $\lambda_i - \lambda_{i+1} \neq 1$,

- $\lambda_i - \lambda_{i+2} \geq 3$,

- $(\lambda_{l(\lambda)-2}, \lambda_{l(\lambda)-1}, \lambda_{l(\lambda)}) \neq (4, 1, 1)$,

- $\lambda_i - \lambda_{i+2} = 3 \implies \lambda_i \neq \lambda_{i+1}$,

- $\lambda_i - \lambda_{i+2} = 3$ and $2 \nmid \lambda_i \implies \lambda_{i+1} \neq \lambda_{i+2}$.

- $\lambda_i - \lambda_{i+2} = 4$ and $2 \nmid \lambda_i \implies \lambda_i \neq \lambda_{i+1}$,

- *Consider the first differences*

$$\Delta\lambda := (\lambda_1 - \lambda_2, \lambda_2 - \lambda_3, \ldots, \lambda_{l-1} - \lambda_l).$$

None of the following subwords are permitted in $\Delta\lambda$:

$$(3, 3, 0), (3, 2, 3, 0), (3, 2, 2, 3, 0), \ldots, (3, 2, 2, 2, 2, \ldots, 2, 3, 0),$$

if the smallest part is at least 5.

Finally, we give the conjectured identity associated with the $(0, 2)$-module.

Conjecture 5.7 (Nandi). *The number of partitions of* n *into parts* $\equiv \pm 2, \pm 5, \pm 6 \pmod{14}$ *equals the number of partitions* $\lambda = (\lambda_1, \lambda_2, \ldots, \lambda_l)$ *of* n *where*

- $m_1(\lambda) = m_3(\lambda) = 0$,

- $m_2(\lambda) \leq 1$,

- $\lambda_i - \lambda_{i+1} \neq 1$,

- $\lambda_i - \lambda_{i+2} \geq 3$,

- $\lambda_i - \lambda_{i+2} = 3 \implies \lambda_i \neq \lambda_{i+1}$,

- $\lambda_i - \lambda_{i+2} = 3$ and $2 \nmid \lambda_i \implies \lambda_{i+1} \neq \lambda_{i+2}$.

- $\lambda_i - \lambda_{i+2} = 4$ and $2 \nmid \lambda_i \implies \lambda_i \neq \lambda_{i+1}$,

- λ *does not contain any of the following subpartitions:*

$$(5, 2), \quad (7, 4, 2), \quad (9, 6, 4, 2), \quad (11, 8, 6, 4, 2), \ldots,$$

- *Consider the first differences*

$$\Delta\lambda := (\lambda_1 - \lambda_2, \lambda_2 - \lambda_3, \ldots, \lambda_{l-1} - \lambda_l).$$

None of the following subwords are permitted in $\Delta\lambda$:

$$(3, 3, 0), (3, 2, 3, 0), (3, 2, 2, 3, 0), \ldots, (3, 2, 2, 2, 2, \ldots, 2, 3, 0),$$

if the least part is at least 5.

Using a sophisticated computer search, Shashank Kanade and Matthew Russell conjectured the following partition identities [KR15] that might be associated with the standard modules of the Lie algebra $D_4^{(3)}$.

Conjecture 5.8 (Kanade and Russell). *Let $KR_1(n)$ denote the number of partitions of n into parts congruent to $\pm 1, \pm 3$ (mod 9). Let $KR_1^*(n)$ denote the number of partitions λ of n such that*

- $\lambda_i - \lambda_{i+2} \geqq 3$,

- $\lambda_i - \lambda_{i+1} \leqq 1 \implies \lambda_i + \lambda_{i+2} \equiv 0$ (mod 3).

Then $KR_1(n) = KR_1^*(n)$ for all n.

Conjecture 5.9 (Kanade and Russell). *Let $KR_2(n)$ denote the number of partitions of n into parts congruent to $\pm 2, \pm 3$ (mod 9). Let $KR_2^*(n)$ denote the number of partitions λ of n such that*

- $m_1(\lambda) = 0$,

- $\lambda_i - \lambda_{i+2} \geqq 3$,

- $\lambda_i - \lambda_{i+1} \leqq 1 \implies \lambda_i + \lambda_{i+2} \equiv 0$ (mod 3).

Then $KR_2(n) = KR_2^*(n)$ for all n.

Conjecture 5.10 (Kanade and Russell). *Let $KR_3(n)$ denote the number of partitions of n into parts congruent to $\pm 3, \pm 4$ (mod 9). Let $KR_3^*(n)$ denote the number of partitions λ of n such that*

- $m_1(\lambda) = m_2(\lambda) = 0$,

- $\lambda_i - \lambda_{i+2} \geqq 3$,

- $\lambda_i - \lambda_{i+1} \leqq 1 \implies \lambda_i + \lambda_{i+2} \equiv 0$ (mod 3).

Then $KR_3(n) = KR_3^*(n)$ for all n.

Remark 5.11. For $i = 1, 2, 3$, the generating functions of $KR_i(n)$ are related to the principal characters of the three inequivalent level 3 standard modules of $D_4^{(3)}$, as presented in Table 5.1.

Closely related to the preceding three conjectures, are two additional conjectures involving asymmetric residue classes modulo 9.

Conjecture 5.12 (Kanade and Russell). *Let $KR_4(n)$ denote the number of partitions of n into parts congruent to $2, 3, 5, 8$ (mod 9). Let $KR_4^*(n)$ denote the number of partitions λ of n such that*

- $m_1(\lambda) = 0$,

- $\lambda_i - \lambda_{i+2} \geqq 3$,

module	principal character
$(1,1,0)$	$(q, q^3, q^6, q^8; q^9)_\infty^{-1}$
$(3,0,0)$	$(q^2, q^3, q^6, q^7; q^9)_\infty^{-1}$
$(0,0,1)$	$(q^3, q^4, q^5, q^6; q^9)_\infty^{-1}$

TABLE 5.1: Principal characters of the level 3 standard modules of $D_4^{(3)}$.

- $\lambda_i - \lambda_{i+1} \leqq 1 \implies \lambda_i + \lambda_{i+2} \equiv 2 \pmod 3$.

Then $KR_4(n) = KR_4^*(n)$ for all n.

Conjecture 5.13 (Kanade and Russell). *Let $KR_5(n)$ denote the number of partitions of n into parts congruent to $1, 4, 6, 7 \pmod 9$. Let $KR_5^*(n)$ denote the number of partitions λ of n such that*

- $m_2(\lambda) \leqq 1$,

- $\lambda_i - \lambda_{i+2} \geqq 3$,

- $\lambda_i - \lambda_{i+1} \leqq 1 \implies \lambda_i + \lambda_{i+2} \equiv 1 \pmod 3$.

Then $KR_5(n) = KR_5^*(n)$ for all n.

Kanade and Russell conjectured two further asymmetric partition identities associated with the modulus 12.

Conjecture 5.14 (Kanade and Russell). *Let $KR_6(n)$ denote the number of partitions of n into parts congruent to $1, 3, 4, 6, 7, 10, 11 \pmod{12}$. Let $KR_6^*(n)$ denote the number of partitions λ of n such that*

- $m_1(\lambda) \leqq 1$,

- $\lambda_i - \lambda_{i+3} \geqq 3$,

- $\lambda_i - \lambda_{i+2} \leqq 1 \implies \lambda_i + \lambda_{i+1} + \lambda_{i+2} \equiv 1 \pmod 3$.

Then $KR_6(n) = KR_6^*(n)$ for all n.

Conjecture 5.15 (Kanade and Russell). *Let $KR_7(n)$ denote the number of partitions of n into parts congruent to $2, 3, 5, 6, 7, 8, 11 \pmod{12}$. Let $KR_7^*(n)$ denote the number of partitions λ of n such that*

- $m_1(\lambda) = 0$,

- $\lambda_i - \lambda_{i+3} \geqq 3$,

- $\lambda_i - \lambda_{i+2} \leqq 1 \implies \lambda_i + \lambda_{i+1} + \lambda_{i+2} \equiv 2 \pmod 3$.

Then $KR_7(n) = KR_7^*(n)$ for all n.

It should be noted that the above conjectures have been verified for all $n \leq 500$, far exceeding MacMahon's criterion of virtual certainty (i.e. $n \leq 89$). These conjectures are, as of this writing, among the most important open problems in partition theory, and have thus far defied proof even by the most experienced practitioners in the field.

The reason that the "motivated proofs" of the type introduced in this chapter are of particular interest to those who are interested in the structure of affine Kac–Moody Lie algebras is that the principal characters of the standard modules of these Lie algebras are known to equal specific infinite products. Thus, informally, we may say that the "product sides" are all known. That these principal characters may also equal certain multisums, or equivalently, generate partitions that satisfy certain (possibly very complicated) difference and initial conditions, is only known in a limited number of particular special cases.

Jim Lepowsky was attracted to the concept of the Andrews–Baxter "motivated" proof because these proofs *only* rely on knowledge of the product side at the beginning, and lead one to discover the corresponding sum side. Thus, it is the long-term goal that by producing and studying motivated proofs that start only with the product side of a Rogers–Ramanujan type identity, we will gain important insights into the structure of the Kac–Moody Lie algebras.

5.2.3 q-Product Formulæ for Principal Characters of Standard Modules of Selected Kac–Moody Lie Algebras

In the late 1990s, Maegan Bos developed a Maple program that explicitly calculates the principal character of a given standard module of a given Kac–Moody Lie algebra. In [Bos03], Bos explains the program and presents a number of interesting related results. For related background and definitions of the technical terms used, see [Kac90] and [Car05].

The following product formulæ are "known" in the sense that they could be constructed from more general results in the literature. However, I have not found them in an easily recognizable form, ready for use by a q-series practitioner, and so I include them below.

5.2.3.1 $A_n^{(1)}$

Let n be a positive integer, and let $s_0, s_1, \ldots, s_n$ be nonnegative integers. The Coxeter number of $A_n^{(1)}$ is $h = n + 1$ and the level ℓ of the standard module $(s_0, s_1, \ldots, s_n)$ is given by $\ell = \sum_{i=0}^{n} s_i$.

Let G denote the Dynkin diagram of $A_n^{(1)}$ viewed as a graph with vertices labeled by the roots $\alpha_0, \alpha_1, \ldots, \alpha_n$. Let $S_{j,L}$ denote a set $\{s_{j_1}, s_{j_2}, \ldots, s_{j_L}\}$, where $\{\alpha_{j_1}, \alpha_{j_2}, \ldots, \alpha_{j_L}\}$ is the set of vertices of a connected subgraph of G with L vertices. Let $\mathfrak{G}_L$ be the collection of all such $S_{j,L}$.

The principal character of the standard $A_n^{(1)}$ module $(s_0, s_1, \ldots, s_n)$ is given by

$$\frac{(q^M; q^M)_\infty^n}{(q; q)_\infty^n} \prod_{L=1}^{n} \prod_{S_{j,L} \in \mathfrak{G}_L} (q^{L + \sum_{k=1}^{L} s_{j_k}}; q^M)_\infty$$

where $M = h + \ell$.

5.2.3.2 $A_2^{(2)}$

Let s_0 and s_1 denote nonnegative integers. The level ℓ of the standard module (s_0, s_1) is $\ell = s_0 + 2s_1$, and the Coxeter number is 3.

The principal character of the (s_0, s_1) module is given by

$$(q^{s_1+1}; q^M)_\infty (q^{s_0+s_1+2}; q^M)_\infty (q^{s_0+1}; q^{2M})_\infty (q^{s_0+4s_1+5}; q^{2M})_\infty \frac{(q^M; q^M)_\infty}{(q; q)_\infty}$$

where $M = \ell + h = s_0 + 2s_1 + 3$.

5.2.3.3 $C_n^{(1)}$

Let $n \geq 2$ and let $s_0, s_1, \ldots, s_n$ be nonnegative integers. The Coxeter number of $C_n^{(1)}$ is $h = 2n$ and the level ℓ of the standard module $(s_0, s_1, \ldots, s_n)$ is given by $\ell = \sum_{i=0}^{n} s_i$.

The principal character of the standard $C_n^{(1)}$ module $(s_0, s_1, \ldots, s_n)$ is given by

$$\prod_{L=1}^{n} \prod_{i=0}^{n-L+1} (q^{L + \sum_{j=i}^{i+L-1} s_j}; q^M)_\infty (q^{M - (L + \sum_{j=i}^{i+L-1} s_j)}; q^M)_\infty$$

$$\times \prod_{L=1}^{n} \prod_{p=1}^{\lfloor \frac{L}{2} \rfloor} (q^{L + 2\sum_{j=0}^{p-1} s_j + \sum_{j=p}^{L-2p+1} s_j}; q^M)_\infty (q^{L + 2\sum_{j=0}^{p-1} s_{n-j} + \sum_{j=p}^{L-2p+1} s_{n-j}}; q^M)_\infty$$

$$\times \prod_{L=1}^{n} \prod_{p=1}^{\lfloor \frac{L}{2} \rfloor} \left[(q^{M - (L + 2\sum_{j=0}^{p-1} s_j + \sum_{j=p}^{L-2p+1} s_j)}; q^M)_\infty \right.$$

$$\left. \times (q^{M - (L + 2\sum_{j=0}^{p-1} s_{n-j} + \sum_{j=p}^{L-2p+1} s_{n-j})}; q^M)_\infty \right]$$

$$\times (q^{M/2}; q^M)_\infty (q^M; q^M)_\infty^n (q; q)_\infty^{-n},$$

where $M = h + 2\ell + 2$.

5.2.3.4　$D_n^{(1)}$

Let $n \geq 4$, and let $s_0, s_1, \ldots, s_n$ be nonnegative integers. The Coxeter number of $D_n^{(1)}$ is $h = 2n - 2$ and the level ℓ of the standard module $(s_0, s_1, \ldots, s_n)$ is given by $\ell = s_0 + s_1 + \left(2 \sum_{i=2}^{n-2} s_i\right) + s_{n-1} + s_n$.

Let G denote the Dynkin diagram of $D_n^{(1)}$ viewed as a graph with vertices labeled by the roots $\alpha_0, \alpha_1, \ldots, \alpha_n$. Let $S_{j,L}$ denote a set $\{s_{j_1}, s_{j_2}, \ldots, s_{j_L}\}$, where $\{\alpha_{j_1}, \alpha_{j_2}, \ldots, \alpha_{j_L}\}$ is the set of vertices of a connected subgraph of G with L vertices. Let $\mathfrak{S}_L$ be the collection of all such $S_{j,L}$.

The principal character of the standard $D_n^{(1)}$ module $(s_0, s_1, \ldots, s_n)$ is given by

$$\frac{(q^M; q^M)_\infty^n}{(q; q)_\infty^n} \prod_{L=1}^{n-2} \prod_{S_{j,L} \in \mathfrak{S}_L} (q^{L + \sum_{k=1}^{L} s_{j_k}}; q^M)_\infty (q^{M - (L + \sum_{k=1}^{L} s_{j_k})}; q^M)_\infty$$

$$\times \prod_{S_{j,n-1} \in \mathfrak{S}_{n-1}} (q^{(n-1) + \sum_{k=1}^{n-1} s_{j_k}}; q^M)_\infty$$

where $M = h + \ell$.

Remark 5.16. The material in this chapter has been developed under the mentorship of Jim Lepowsky, to whom the author owes many thanks. The author also wishes to point out that the interested reader may wish to consult Victor Kac's book *Infinite Dimensional Lie Algebras* [Kac90] for a possibly different perspective on related material.

Chapter 6

But wait ... there's more!

Indeed, there is much, much more to say about the Rogers–Ramanujan identities than can reasonably be covered in a volume of this size. In this chapter, we will point the interested reader in the direction of a number of additional topics.

6.1 Partition Bijections

6.1.1 Euler's Partition Identity

While many partition identities can be proved (and often proved rather efficiently) using arguments involving generating functions, there is something particularly satisfying in knowing a direct bijection between two equinumerous restricted classes of partitions. As we have seen earlier, Euler's partition theorem, Theorem 1.8, that the number of partitions of n into odd parts equals the number of partitions of n into distinct parts, is an immediate consequence of the equality of the infinite products

$$\frac{1}{(q;q^2)_\infty} = (-q;q)_\infty,$$

as shown by Euler in the 1740s. The first bijective proof of this theorem was given by the English mathematician and astronomer J. W. L. Glaisher in 1883 [Gla83].

Glaisher's proof is simple and elegant, and we present it now.

Bijective proof of Theorem 1.8.

- Suppose $\lambda = \langle 1^{m_1} 3^{m_3} 5^{m_5} \cdots \rangle$ is a partition of n into all odd parts.

- For each odd i, find the binary (base 2) expansion of m_i,

$$m_i = \sum_{j \geq 0} a_{ij} 2^j,$$

  where each $a_{ij} \in \{0, 1\}$ and, of course, only finitely many of the a_{ij} equal 1.

- Now observe that

$$
\begin{aligned}
n =& (\cdots + a_{13}2^3 + a_{12}2^2 + a_{11}2^1 + a_{10}) \cdot 1 \\
& (\cdots + a_{33}2^3 + a_{32}2^2 + a_{31}2^1 + a_{30}) \cdot 3 \\
& (\cdots + a_{53}2^3 + a_{52}2^2 + a_{51}2^1 + a_{50}) \cdot 5 \\
& \vdots
\end{aligned}
$$
$$\qquad\qquad\qquad\qquad\qquad\qquad\qquad\qquad\qquad (6.1)$$

$$=a_{10} \cdot 1 + a_{11} \cdot 2 + a_{30} \cdot 3 + a_{12} \cdot 4 + a_{50} \cdot 5 + \cdots, \qquad (6.2)$$

  where the last equality simply follows from the distributive law of multiplication over addition.

- By the construction, for each integer k, there is exactly one coëfficient a_{ij}, because each integer k can be written uniquely as $k = 2^\alpha \omega$, where $\alpha = \alpha(k)$ is a nonnegative integer and ω is odd. (This follows immediately from the fundamental theorem of arithmetic.) Simply take ω equal to the largest odd divisor of k, and then $\alpha = \log_2(k/\omega)$. Thus the coëfficient of k in (6.2) is $a_{\omega(k),\alpha(k)}$.

- From (6.2), form the partition

$$\pi = \langle 1^{a_{10}} 2^{a_{11}} 3^{a_{30}} 4^{a_{12}} \cdots k^{a_{\omega\alpha}} \cdots \rangle.$$

- Clearly π is a partition of n into distinct parts, since each $a_{\omega\alpha} \in \{0, 1\}$.

- The steps are reversible, and so the indicated construction forms a bijection between the set of partitions of n into odd parts and the set of partitions of n into distinct parts.

$\square$

6.1.2   Garsia–Milne Involution Principle

Paul Erdős, the legendary itinerant mathematician, enjoyed offering prizes (between \$10 and \$3000) for the published solutions to particular problems he posed. George Andrews has emulated this practice of Erdős on a number of occasions, including one time when he offered \$50 for a bijective proof of the first Rogers–Ramanujan identity. Adriano Garsia and Steve Milne rose to the challenge and invented a new method—the involution principle—to construct a bijection between the set of partitions of n into parts congruent to ± 1 (mod 5) and the set of partitions of n where no consecutive integers appear as parts. This bijection is no straight forward business, however. Their proof takes up the majority of a 51-page paper [GM81]. The proof is complicated due to the details that arise, but the idea behind the involution principle itself is actually rather simple, and is indeed very elegant. Bressoud and Zeilberger used the involution principle to produce a short (but still difficult) Rogers–Ramanujan bijection [BZ82].

Intrigued beginners are advised to first study the involution principle proof of Euler's partition identity presented by Andrews in [And86, pp. 60–62].

It is interesting to note that the bijection obtained by the involution principle applied to Euler's partition identity turns out to be the same as that of Glaisher, albeit via a much more complicated route. Indeed, when the involution principle is applied to a classical partition theorem for which a (straightforward) bijective proof is known, invariably the classical bijection is recovered. This phenomenon was studied by Jeff Remmel [Rem82], Basil Gordon [Gor83], and Kathy O'Hara [O'H88].

For a combinatorial, but not *entirely* bijective, proof of the first Rogers–Ramanujan identity, see the paper by Cilanne Boulet and Igor Pak [BP06]. Pak also has a wonderful survey article on partition bijections [Pak06].

6.1.3 Exercise

Generalize Euler's partition theorem to the following.

Theorem 6.1. *Let d and n denote positive integers. The number of partitions of n into parts not divisible by d equals the number of partitions of n where no integer appears as a part more than d − 1 times.*

Give a generating function proof and a bijective proof.

6.2 Rogers–Ramanujan Continued Fraction

In the same paper [Rog94] where the Rogers–Ramanujan identities first appeared, Rogers also presented what is now known as the Rogers–Ramanujan continued fraction:

$$R(q) := \cfrac{q^{1/5}}{1 + \cfrac{q}{1 + \cfrac{q^2}{1 + \cfrac{q^3}{1 + \cdots}}}}. \tag{6.3}$$

Rogers showed that $R(q)$ is, up to a factor, equal to the ratio of the two Rogers–Ramanujan functions, i.e.

$$R(q) = q^{1/5}\frac{H(q)}{G(q)} = q^{1/5}\frac{f(-q,-q^4)}{f(-q^2,-q^3)},$$

where

$$G(q) := \sum_{n\geq 0}\frac{q^{n^2}}{(q;q)_n},$$

and

$$H(q) := \sum_{n\geq 0}\frac{q^{n^2+n}}{(q;q)_n}.$$

(The $q^{1/5}$ factor is tacked on to make $R(q)$ a modular form.)

Ramanujan independently rediscovered (6.3), and in fact presented corollaries of (6.3) in his first letter to Hardy [BR95, pp. 21–30], dated January 16, 1913. Specifically, Ramanuajan included the results:

$$\left(\sqrt{\frac{5+\sqrt{5}}{2}} - \frac{\sqrt{5}+1}{2}\right)e^{2\pi/5} = \cfrac{1}{1+\cfrac{e^{-2\pi}}{1+\cfrac{e^{-4\pi}}{1+\cfrac{e^{-6\pi}}{1+\cdots}}}} \qquad (6.4)$$

and

$$\left(\sqrt{\frac{5-\sqrt{5}}{2}} - \frac{\sqrt{5}-1}{2}\right)e^{2\pi/5} = \cfrac{1}{1+\cfrac{e^{-\pi}}{1+\cfrac{e^{-2\pi}}{1+\cfrac{e^{-3\pi}}{1+\cdots}}}}. \qquad (6.5)$$

Of (6.4) and (6.5), Hardy stated in a lecture [Har37a] delivered at the Harvard Tercentenary Conference of Arts and Sciences on August 31, 1936:

> [These formulæ] defeated me completely; I had never seen anything in the least like them before. A single look at them is enough to show that they could only be written down by a mathematician of the highest class. They must be true because, if they were not true, no one would have had the imagination to invent them. Finally (you must remember that I knew nothing whatever about Ramanujan, and had to think of every possibility), the writer must be completely honest, because great mathematicians are commoner than thieves or humbugs of such incredible skill.

The first five chapters of Andrews and Berndt's book *Ramanujan's Lost Notebook, Part I* [AB05] are devoted to the Rogers–Ramanujan continued fraction and related results, and the intrigued reader is urged to look there. Among the results discussed there are Ramanujan's surprising and beautiful relations:

$$\frac{1}{R(q)} - 1 - R(q) = \frac{(q^{1/5}; q^{1/5})_\infty}{q^{1/5}(q^5; q^5)_\infty}, \tag{6.6}$$

$$\frac{1}{R^5(q)} - 11 - R^5(q) = \frac{(q; q)_\infty^6}{q(q^5; q^5)_\infty^6}. \tag{6.7}$$

Like the Rogers–Ramanujan identities, the Rogers–Ramanujan continued fraction is one of many similar results.

A modulus 6 analog of (6.3) is known as the Ramanujan cubic continued fraction:

$$\cfrac{1}{1 + \cfrac{q + q^2}{1 + \cfrac{q^2 + q^4}{1 + \cfrac{q^3 + q^6}{1 + \cdots}}}} = \frac{f(-q, -q^5)}{f(-q^3, -q^3)}; \tag{6.8}$$

see, e.g., [AB05, §3.3, pp. 94–100].

Following Basil Gordon's dictum "where there is a 5, there is an 8," there is also a modulus 8 analog of (6.3), called the Ramanujan–Göllnitz–Gordon continued fraction. It was recorded by Ramanujan in the lost notebook, and independently rediscovered by both Göllnitz [Göl67] and Gordon [Gor65].

$$\cfrac{1}{1 + \cfrac{q + q^2}{1 + \cfrac{q^4}{1 + \cfrac{q^3 + q^6}{1 + \cdots}}}} = \frac{f(-q, -q^7)}{f(-q^3, -q^5)}; \tag{6.9}$$

see, e.g., [AB05, p. 154, Eq. (6.2.38)] and the article by Heng Huat Chan and Sen-Shan Huang [CH97].

6.3 Statistical Mechanics

The Rogers–Ramanujan identities and identities of similar type first appeared in physics in Rodney Baxter's exact solution to the hard hexagon model in statistical mechanics. George Andrews has mentioned in numerous lectures over the years that "the Rogers–Ramanujan identities are intimately linked with the behavior of liquid helium on a graphite plate." The author has been unable to locate this statement in Baxter's writings (surely it must be there somewhere), but see the paper coauthored by Baxter's colleague at Australian National University Murray Batchelor, and Angela Foerster of Chongqing University [BF16] for a confirmation in the literature.

The name "hard hexagon model" arises from the following scenario: consider the integer (lattice) points in the plane as possible locations where a particle may reside. Suppose that there is a particle located at the point (j, k) (where j and k are integers). The model then declares that no particle may reside at any of the six points $(j-1, k+1)$, $(j, k+1)$, $(j-1, k)$, $(j+1, k)$, $(j, k-1)$, $(j+1, k-1)$. These six forbidden points are then the vertices of a certain hexagon. An equivalent problem is therefore packing such hexagons centered on integer points in such a way that the hexagons do not overlap. Hence, the name "hard hexagon model."

A specific arrangement of particles is called a state s of the system. Associated with each state s is the total number of particles $n(s)$ and energy $E(s)$. Next, there is a *partition function*

$$Z = \sum_{\text{all } s} e^{-\beta E(s) - mu(s)},$$

where the use of the word "partition" is not the same as elsewhere in this book. In order to evaluate Z and other quantities, Baxter utilizes *corner transfer matrices* (Baxter [Bax82, Chapter 13]). The Rogers–Ramanujan series naturally arise from the eigenvalues of the associated corner transfer matrix. To determine the phase transitions associated with the physical system, it is necessary to estimate the asymptotics of the associated Rogers–Ramanujan type series as $q \to 1$ from the left with extreme accuracy. These accurate asymptotics are readily obtained by appealing to the infinite product side of the corresponding Rogers–Ramanujan type identity, where modular transformation theory is most naturally applied.

The curious reader is advised to consult Chapter 8 of Andrews' q-series monograph [And86] for a more detailed overview. The full story is contained in Baxter's classic book *Exactly Solved Models in Statistical Mechanics* [Bax82], which had been out of print for many years, but was republished by Dover in 2008. Indeed, Chapter 13 of [Bax82] is devoted to corner transfer matrices and a detailed explanation of the solution of the hard hexagon model. See also Baxter's papers [Bax80, Bax81]. The hard hexagon model was greatly generalized by Andrews, Baxter, and Forrester [ABF84], Forrester and Baxter [FB85], and Jimbo and Miwa [JM85].

Baxter was awarded the 1980 Boltzman Medal for his work on the hard hexagon model. In 1998, Andrews was awarded an honorary doctorate in physics from the University of Parma, for his work in statistical mechanics, much of which was done in collaboration with Baxter.

Additional work on models in statistical mechanics and Rogers–Ramanujan type identities continued throughout the 1990s and into the

twenty-first century, including papers by various subsets of the following practitioners: Andrews, Baxter, Alexander Berkovich, Barry McCoy,
William Orrick, Paul Pearce, Anne Schilling, and Ole Warnaar. See, e.g.,
[AB87, BP83, BP84, BM96, BM97, BM98, BMO96, BMS98, SW98, War01b],
each of which in turn points to other relevant references.

6.4 Loxton's Special Values of the Dilogarithm Function

In a paper published in 1984 [Lox84], John Loxton studies special values
of the dilogarithm function, which is defined as

$$\mathrm{Li}_2(z) := \sum_{n \geq 1} \frac{z^n}{n^2} = - \int_0^z \frac{\log(1-t)}{t} \, dt$$

for suitable values of z, and derived a large number of known Rogers–
Ramanujan type identities, and a few new ones, including

$$\sum_{n \geq 0} \frac{q^{n(n+1)}(-1; q^3)_n}{(-1; q)_n (q; q)_{2n}} = \frac{(q, q^8, q^9; q^9)_\infty (q^7, q^{11}; q^{18})_\infty}{(q; q)_\infty} \qquad (6.10)$$

and

$$\sum_{n \geq 0} \frac{q^{n^2}(-1; q^3)_n}{(-1; q)_n (q; q)_{2n}} = \frac{(q^2, q^7, q^9; q^9)_\infty (q^5, q^{13}; q^{18})_\infty}{(q; q)_\infty}. \qquad (6.11)$$

Unaware of Loxton's work at the time, McLaughlin and Sills rediscovered (6.10) and (6.11), along with two partner identities not found by Loxton [MS08]:

$$\sum_{n \geq 0} \frac{q^{n(n+1)}(-q^3; q^3)_n}{(-q; q)_n (q; q)_{2n+1}} = \frac{(q^3, q^6, q^9; q^9)_\infty (q^3, q^{15}; q^{18})_\infty}{(q; q)_\infty} \qquad (6.12)$$

and

$$\sum_{n \geq 0} \frac{q^{n(n+2)}(-q^3; q^3)_n}{(q^2; q^2)_n (q^{n+2}; q)_{n+1}} = \frac{(q^4, q^5, q^9; q^9)_\infty (q, q^{17}; q^{18})_\infty}{(q; q)_\infty}. \qquad (6.13)$$

All of (6.10)–(6.13) can be proved by inserting certain Bailey pairs into the
appropriate limiting case of Bailey's lemma (2.39). Indeed, Slater had all the
necessary tools in hand to have discovered all four of these identities back in
the late 1940s, but simply did not.

Also, it should be noted that the product sides of (6.10)–(6.13) correspond
to the principal characters of the standard $A_2^{(2)}$ level 6 modules $(6,0)$, $(4,1)$,

$(2,2)$, and $(0,3)$ respectively. Whether the sum sides contain any vertex operator theoretic information is unknown at this time.

Exercises

1. Let

$$\beta_n = \frac{(-1;q^3)_n}{(q;q)_{2n}(-1;q)_n}$$

and

$$\alpha_n = \begin{cases} 1 & \text{if } n = 0 \\ -q^{(9r^2+9r+2)/2} & \text{if } n = 3r+1 \\ -q^{(9r^2-9r-2)/2} & \text{if } n = 3r-1 \\ q^{3(3r-1)/2}(1+q^{3r}) & \text{if } n = 3r > 0 \end{cases}.$$

Prove that (α_n, β_n) forms a Bailey pair.

2. Deduce (6.11).

3. Deduce the identity

$$\sum_{n\geq 0} \frac{q^{n^2}(-q;q^2)_n(-1;q^6)_n}{(q^2;q^2)_{2n}(-1;q^2)_n} = \frac{(q^3,q^9,q^{12};q^{12})_\infty (q^6,q^{18};q^{24})_\infty}{\psi(-q)}. \quad (6.14)$$

6.5   Hall–Littlewood Polynomials

In [Ste90], John Stembridge used the theory of Hall–Littlewood functions (see Ian Macdonald's book [Mac99] for an introduction) to derive Rogers–Ramanujan type identities. In addition to obtaining the Rogers–Ramanujan identities themselves, he derived a number of families of identities. He presented these identities in a form different from what we have seen previously in this book; in particular one side of each identity involves a sum indexed by certain partitions, rather than by the nonnegative integers. For example, Stembridge found [Ste90, p. 477, Corollary 1.5, Eq. (a)]: for all positive integers k,

$$\sum_{\lambda \in \mathscr{P}_k} \frac{q^{n(\lambda)+|\lambda|}}{(q;q)_\lambda} = \frac{f(-q,-q^{k+1})}{\varphi(-q)}, \quad (6.15)$$

where $\mathscr{P}_k$ denotes the set of all partitions of length at most k,

$$n(\lambda) := \sum_{i=1}^{\ell(\lambda)} \binom{\lambda_i}{2},$$

and

$$(q;q)_\lambda := \prod_{i=1}^{\lambda_1'} (q;q)_{m_i(\lambda')} = \prod_{i=1}^{\ell(\lambda)} (q;q)_{\lambda_i - \lambda_{i+1}},$$

where we follow the convention that $\lambda_{\ell(\lambda)+1} = 0$.

In [GOW16], Michael Griffin, Ken Ono, and Ole Warnaar show that the Andrews–Gordon identity is a special case of a doubly-infinite family of identities connected with the Kac–Moody algebra $A_{2n}^{(2)}$. (This connection is *not* the same as the connection using the standard modules of $A_2^{(2)}$ via vertex operators that gives rise to the Capparelli and Nandi partition identities discussed earlier.)

The sum sides of the Griffin–Ono–Warnaar identities are expressed in terms of Hall–Littlewood polynomials

$$P_\lambda(x; q) := \frac{1}{v_\lambda(q)} \sum_{w \in \mathfrak{S}_n} w \left(x^\lambda \prod_{i<j} \frac{x_i - qx_j}{x_i - x_j} \right),$$

where the symmetric group $\mathfrak{S}_n$ acts on $x = (x_1, x_2, \ldots, x_n)$ by permuting the x_i; for a partition λ,

$$x^\lambda := \prod_{i=1}^n x_i^{\lambda_i},$$

$$v_\lambda(q) := \prod_{i=0}^n \frac{(q; q)_{m_i(\lambda)}}{(1-q)^{m_i(\lambda)}},$$

($m_0(\lambda) := n - \ell(\lambda)$; for $i > 0$, $m_i(\lambda)$ denotes the multiplicity of i in λ, as usual).

For instance, they found [GOW16, Theorem 1.1], among many beautiful general results, that for positive integers m and n,

$$\sum_{\lambda:\lambda_1 \leq m} q^{|\lambda|} P_{2\lambda}(1, q, q^2, \ldots; q^{2n-1})$$

$$= \frac{1}{(q; q)_\infty^n (q^{2n+2m+1}; q^{2n+2m+1})_\infty^{n(n-1)}} \prod_{i=1}^n f(-q^{i+m}, -q^{m+2n+1-i})$$

$$\times \prod_{1 \leq i < j \leq n} f(-q^{j-i}, -q^{2n+2m+1-j+i}) f(-q^{i+j}, -q^{2n+2m+1-i-j})$$

$$= \frac{1}{(q; q)_\infty^m (q^{2n+2m+1}; q^{2n+2m+1})_\infty^{m(m-1)}} \prod_{i=1}^m f(-q^{i+1}, -q^{2m+2n-i})$$

$$\times \prod_{1 \leq i < j \leq n} f(-q^{j-i}, -q^{2n+2m+1-j+i}) f(-q^{i+j}, -q^{2n+2m+1-i-j}), \quad (6.16)$$

and

$$\sum_{\lambda:\lambda_1 \leq m} q^{2|\lambda|} P_{2\lambda}(1, q, q^2, \ldots; q^{2n-1})$$

$$= \frac{1}{(q;q)_\infty^n (q^{2n+2m+1};q^{2n+2m+1})_\infty^{n(n-1)}} \prod_{i=1}^{n} f(-q^i, -q^{2m+2n+1-i})$$

$$\times \prod_{1 \leq i < j \leq n} f(-q^{j-i}, -q^{2n+2m+1-j+i}) f(-q^{i+j}, -q^{2n+2m+1-i-j})$$

$$= \frac{1}{(q;q)_\infty^m (q^{2n+2m+1};q^{2n+2m+1})_\infty^{m(m-1)}} \prod_{i=1}^{m} f(-q^i, -q^{2m+2n+1-i})$$

$$\times \prod_{1 \leq i < j \leq n} f(-q^{j-i}, -q^{2n+2m+1-j+i}) f(-q^{i+j}, -q^{2n+2m+1-i-j}). \quad (6.17)$$

Note that when $m = n = 1$, the preceding result gives the Rogers–Ramanujan identities, as the sum is over the empty partition and the partition $\lambda = \langle 1^n \rangle$. For $n = m = 2$, we get a form Eq. (2.26), Dyson's personal favorite among the various identities of Rogers–Ramanujan type he discovered as a youngster assigned by Hardy to referee Bailey's paper [Bai47]. (See Section 2.3.1.)

6.6   "m-Versions"

We are by now well acquainted with the two Rogers–Ramanujan identities,

$$\sum_{n \geq 0} \frac{q^{n^2}}{(q;q)_n} = \frac{1}{(q, q^4; q^5)_\infty}$$

and

$$\sum_{n \geq 0} \frac{q^{n^2+n}}{(q;q)_n} = \frac{1}{(q^2, q^3; q^5)_\infty}.$$

It therefore seems natural to consider the series

$$\sum_{n \geq 0} \frac{q^{n^2+mn}}{(q;q)_n} \quad (6.18)$$

for values of m beside $m = 0, 1$.

Tina Garrett, Mourad Ismail, and Dennis Stanton [GIS99] proved that for m a nonnegative integer,

$$\sum_{n=0}^{\infty} \frac{q^{n^2+mn}}{(q;q)_n} = \frac{(-1)^m q^{m(m-1)/2} E_{n-2}(q)}{(q, q^4; q^5)_\infty} + \frac{(-1)^m q^{m(m-1)/2} D_{n-1}(q)}{(q^2, q^3; q^5)_\infty}, \quad (6.19)$$

where $D_n(q)$ and $E_n(q)$ are the q-Fibonacci polynomials of Schur defined recursively in (4.4) and (4.17), extended backward using the recurrences to include

$$D_{-1}(q) = 0, \qquad E_{-2}(q) = 1 \qquad E_{-1}(q) = 0.$$

Taking into account the limiting cases of $D_n(q)$ and $E_n(q)$, Equation (6.19) could be written

$$\sum_{n=0}^{\infty} \frac{q^{n^2+mn}}{(q;q)_n} = (-1)^m q^{m(m-1)/2} \left(E_{n-2}(q) D_\infty(q) - D_{n-1}(q) E_\infty(q) \right). \quad (6.20)$$

Leonard Carlitz had earlier found a result similar to (6.19); see [Car59, p. 247, Eqs. (22) and (23)].

We also point out that Nancy Gu and Helmut Prodinger found analogous "m-versions" of many Rogers–Ramanujan type identities in [GP10]. Finally, we note that George Andrews, Arnold Knopfmacher, and Peter Paule further generalized (6.19) in [AKP00] using so-called "q-Engel expansions."

Exercise. Show that (6.18) is the generating function for the number of partitions of n with difference at least 2 between parts, and with all parts greater than or equal to m.

6.7   Knot Theory

In recent years, q-series identities have arisen in topology; specifically in knot theory.

For example, in [GL15], Stavros Garoufalidis and Thang T. Q. Lê prove the multisum q-series to infinite product identity

$$\sum_{i,j,k\geq 0} \frac{(-1)^i q^{i(3i+1)/2+ij+ik+jk+j+k}}{(q)_i(q)_j(q)_k(q)_{i+j}(q)_{i+k}} = \frac{1}{(q;q)_\infty^2}, \quad (6.21)$$

which is related to the 3_1 knot, and the identity

$$\sum_{\substack{i,j,k,l,m\geq 0 \\ i+j=l+m}} \frac{(-1)^{j+l} q^{j^2/2+l^2/2+jk+ik+il+jm+i/2+k+m/2}}{(q)_{j+k}(q)_i(q)_j(q)_k(q)_l(q)_m(q)_{k+l}} = \frac{1}{(q;q)_\infty^3}, \quad (6.22)$$

which is related to the 4_1 knot, where we use the standard abbreviation $(q)_n$ for $(q;q)_n$, in order to save space. For the amphicheiral knot 6_3, they conjectured (and Andrews proved in [And14]) the identity

$$\sum_{\substack{a,b,c,d,e,f\geq 0 \\ a+e\geq b, b+f\geq a}} \frac{(-1)^{a-b+e}q^{F(a,b,c,d,e,f)}}{(q)_a(q)_b(q)_c(q)_{a+c}(q)_d(q)_{a+d}(q)_e(q)_{a-b+e}(q)_{a-b+d+e}(q)_f(q)_{b-a+f}}$$

$$= \frac{1}{(q;q)_\infty^4}, \quad (6.23)$$

where

$$F(a,b,c,d,e,f) := a(3a+1)/2 + b(b+1)/2 + c + ac + d + ad + cd + e(3e+1)/2$$
$$+ 2ae - 2be + de - af + bf + f^2$$

In fact, Andrews [And14, p. 18, Theorem 1] proved three-parameter generalizations of each of (6.21)–(6.23), all of which preserve the multisum-product structure!

6.8   Who Knows What Else?

As of this writing, the Rogers–Ramanujan identities are approaching their 125th birthday, and the centennial of their rise to prominence as an "unproven" conjecture dreamed up by Ramanujan. The past 100 years have witnessed their importance in a variety of mathematical and scientific endeavors. All of us who have had the good fortune to become captivated by these amazing identities have little doubt that over the next century, they will continue to enthrall and inspire new generations of mathematicians and scientists, who will adopt them as beacons for their own work. Long live the Rogers–Ramanujan identities!

Appendices

Appendices

Appendix A

For the rising q-factorial with base q, the abbreviation $(a)_n = (a; q)_n$ is used throughout this appendix. References to Slater's list [Sla52] are simply given in the form "S.n", which is to be interpreted as [Sla52, Eq. (n)]. If S.n is followed by a minus sign, this means that q has been replaced by $-q$; if followed by a "c" this means that a misprint in [Sla52] is being corrected here.

A.1 Nearly Two Hundred Rogers–Ramanujan Type Identities

The subsections are numbered to correspond to the modulus associated with the numerator of the infinite product on the product side of the identities. The author hopes that readers will find this list useful, but there is no suggestion that this appendix is "complete" in any sense. Only single-sum identities where the power of q is a quadratic function of the summation variable are included (thus, e.g., **S. 24** and **S. 26** are excluded).

It is common in the literature to suppress the base in the rising q-factorial notation when the base is q, i.e., to write $(a)_n$ for $(a; q)_n$ and $(a)_\infty$ for $(a; q)_\infty$. We have avoided doing so up to this point in order to avoid possible confusion with ordinary rising factorials. However, to save space in this appendix, this abbreviation will be employed as needed.

As we have throughout the book, we continue to use Ramanujan's notations

$$f(a, b) = (a; ab)_\infty (b; ab)_\infty (ab; ab)_\infty,$$

$$\varphi(q) = f(q, q) = \frac{(-q; -q)_\infty}{(q; -q)_\infty},$$

$$\psi(q) = f(q, q^3) = \frac{(q^2; q^2)_\infty}{(q; q^2)_\infty},$$

$$\chi(q) = \frac{f(-q^2, -q^2)}{(q)_\infty} = (-q; q^2)_\infty,$$

and introduce in this appendix the following notation for instances of the quintuple product identity:

$$Q(w, x) := f(-wx^3, -w^2x^{-3}) + xf(-wx^{-3}, -w^2x^3)$$
$$= (-w/x; w)_\infty (-x; w)_\infty (w; w)_\infty (w/x^2; w^2)_\infty (wx^2; w^2)_\infty.$$

Also, the Rogers–Ramanujan products are

$$G(q) := \frac{f(-q^2, -q^3)}{(q)_\infty} = \frac{1}{(q; q^5)_\infty (q^4; q^5)_\infty},$$

$$H(q) := \frac{f(-q, -q^4)}{(q)_\infty} = \frac{1}{(q^2; q^5)_\infty (q; q^5)_\infty}.$$

In some cases, more than one series expansion is known for a given infinite product. In these cases, the product is not repeated. In other words, if a line begins with an equal sign, this line is to be interpreted as a continuation of the preceding line, although no punctuation marks are employed.

A.1.0   *q*-series Expansions of Constants

$$0 = \sum_{n=0}^{\infty} \frac{(-1)^n q^{n(n-1)/2}}{(q)_n} \qquad \text{(sp. case of Euler's (1.100))} \qquad \text{(A.1)}$$

$$1 = \sum_{n=0}^{\infty} \frac{(-1)^n q^{n^2}}{(q; q^2)_{n+1}} \qquad \text{([Rog17, p. 333 (4)])} \qquad \text{(A.2)}$$

$$1 = \sum_{n=0}^{\infty} \frac{q^{n(n+1)} (q^2; q^2)_{n+1}}{(-q^3; q^3)_{n+1} (q)_n} \qquad \text{([MSZ09, Eq. (2.3)])} \qquad \text{(A.3)}$$

$$2 = \sum_{n=0}^{\infty} \frac{q^{n(n-1)/2}}{(-q)_n} \qquad \text{(A.4)}$$

A.1.1   *q*-series Expansions of Theta Functions and Variants

$$(q)_\infty = \sum_{n=0}^{\infty} (-1)^n q^{n(3n-1)/2} (1 + q^n) \qquad \text{(Euler's Thm. 1.19; S. 1)}$$
$$\text{(A.5)}$$

$$\frac{1}{(q)_\infty} = \sum_{n=0}^{\infty} \frac{q^{n^2}}{(q)_n^2} \qquad \text{(Jacobi, [Jac29, p. 180, Eq. (2)])}$$
$$\text{(A.6)}$$

$$\varphi(-q) = \sum_{n=0}^{\infty} \frac{(-1)^n q^{n(n+1)/2} (q)_n}{(-q)_n}$$

$$([\text{Sta31, p. 805, (3.6)}]; (1.47) \text{ with } a = -c = q) \quad (\text{A.7})$$

$$\frac{1}{\varphi(-q)} = \sum_{n=0}^{\infty} \frac{q^{n(n+1)/2}(-1)_n}{(q)_n^2}$$

$$([\text{Sta31, p. 805, (3.7)}]; (1.47) \text{ with } a = -1, c = q) \quad (\text{A.8})$$

$$\psi(q) = 1 - \sum_{n=1}^{\infty} \frac{(-1)^n q^{n^2+n-1}(1-q)}{(q^2;q^2)_n(1-q^{2n-1})}$$

$$([\text{Sta31, p. 807, (3.14)}]) \quad (\text{A.9})$$

$$\frac{1}{\psi(q)} = \sum_{n=1}^{\infty} \frac{(-1)^n (q;q^2)_n q^{n^2}}{(q^2;q^2)_n^2}$$

$$(\text{Ramanujan [AB09, p. 84, Entry 4.2.6]}) \quad (\text{A.10})$$

$$(-q)_\infty = \sum_{n=0}^{\infty} \frac{q^{n(3n-1)/2}(1+q^{2n})(-q)_{n-1}}{(q)_n}$$

$$([\text{Sta31, p. 809, (3.29)}]) \quad (\text{A.11})$$

A.1.2 Mod 2 Identities

$$\frac{f(-q,-q)}{(q)_\infty} = \sum_{n=0}^{\infty} \frac{(-1)^n q^{n^2}}{(q^2;q^2)_n}$$
$$(\text{sp. case of (1.100); \textbf{S. 3}}) \quad (\text{A.12})$$

$$\frac{f(-q,-q)}{\psi(-q)} = \sum_{n=0}^{\infty} \frac{(-1)^n q^{n^2}(-q;q^2)_n}{(q^4;q^4)_n}$$
$$(\text{Ramanujan [AB09, p. 104, Entry 5.3.6]; \textbf{S. 4}}) \quad (\text{A.13})$$

$$\frac{f(1,q^2)}{\psi(-q)} = \sum_{n=0}^{\infty} \frac{q^{n(n-1)}(-q;q^2)_n}{(q)_{2n}} \qquad \text{sp. case of (1.103)} \quad (\text{A.14})$$

$$= 2\sum_{n=0}^{\infty} \frac{q^{n(n+1)}(-q;q^2)_n}{(q)_{2n+1}} \qquad \text{sp. case of (1.103)} \quad (\text{A.15})$$

A.1.3   Mod 3 Identities

$$\frac{f(-q,-q^2)}{\varphi(-q)} = \sum_{n=0}^{\infty} \frac{q^{n(n+1)/2}}{(q)_n} \text{(sp. case of Euler's (1.100); \textbf{S. 2})} \qquad (A.16)$$

$$\frac{f(-q,-q^2)}{\varphi(-q^2)} = \sum_{n=0}^{\infty} \frac{(-1)^n q^{n(2n+1)}}{(-q;q^2)_{n+1}(q^2;q^2)_n} \qquad\qquad (\textbf{S. 5}) \qquad\qquad (A.17)$$

$$\frac{f(q,q^2)}{(q)_\infty} = \sum_{n=0}^{\infty} \frac{q^{n^2}(-1)_n}{(q)_n(q;q^2)_n} \qquad \text{(Ramanujan [AB09, p. 85, Ent. 4.2.8]; \textbf{S. 6}c)}$$

$$(A.18)$$

$$= \sum_{n=0}^{\infty} \frac{q^{n^2}(-q)_n}{(q)_n(q;q^2)_{n+1}} \qquad \text{(Ramanujan [AB09, p. 86, Entry 4.2.9])}$$

$$(A.19)$$

$$\frac{f(q,q^2)}{\psi(q)} = \sum_{n=0}^{\infty} \frac{q^{2n^2}(q;q^2)_n^2}{(q^2;q^2)_{2n}} \qquad \text{(Ramanujan [AB09, p. 102, Entry 5.3.3])}$$

$$(A.20)$$

$$\frac{f(-q,q^2)}{(q)_\infty} = \sum_{n=0}^{\infty} \frac{q^{n(n+1)}(-1;q^2)_n}{(q)_{2n}}$$

$$\text{(Ramanujan [AB09, p. 86, Ent. 4.2.10]; \textbf{S. 48})} \qquad (A.21)$$

A.1.4   Mod 4 Identities

$$\frac{f(-q,-q^3)}{(q)_\infty} = \sum_{n=0}^{\infty} \frac{q^{n(n+1)}}{(q^2;q^2)_n} \qquad\qquad \text{(sp. case of Euler's (1.100); \textbf{S. 7})}$$

$$(A.22)$$

$$\frac{f(-q^2,-q^2)}{\psi(-q)} = \sum_{n=0}^{\infty} \frac{q^{n^2}(q;q^2)_n}{(q^4;q^4)_n} \qquad\qquad ((1.103) \text{ with } a^2 = -c = q; \textbf{S. 4}-)$$

$$(A.23)$$

$$\frac{f(-q,-q^3)}{\varphi(-q)} = \sum_{n=0}^{\infty} \frac{q^{n(n+1)/2}(-q)_n}{(q)_n} \qquad\qquad ((1.102) \text{ with } a = -q; \textbf{S. 8})$$

$$(A.24)$$

$$= \sum_{n=0}^{\infty} \frac{q^{n(n+1)}(-q;q^2)_n}{(q)_{2n+1}} \qquad ((1.103) \text{ with } -aq = c = q^{3/2}; \textbf{S. 51})$$

$$(A.25)$$

$$\frac{f(-q^2,-q^2)}{\varphi(-q)} = \sum_{n=0}^{\infty} \frac{q^{n(n+1)/2}(-1)_n}{(q)_n}$$

$$((1.102);\ \text{Ramanujan[AB09, p. 38, Entry 1.7.14];}\ \textbf{S. 12})$$
$$(\text{A.26})$$

$$= \sum_{n=0}^{\infty} \frac{q^{n(n+1)/2}(-q)_{n+1}}{(q)_n}$$

$$((1.102)\ \text{with}\ a = -q^2) \qquad\qquad (\text{A.27})$$

$$= \sum_{n=0}^{\infty} \frac{q^{n^2}(-q^2;q^2)_n}{(q)_{2n+1}}$$

$$((1.103);\ \text{Ramanujan [AB09, p. 37, Entry 1.7.13]}) \qquad (\text{A.28})$$

$$= \sum_{n=0}^{\infty} \frac{q^{n^2}(-1;q^2)_n}{(q)_{2n}}$$

$$((1.103)\ \text{with}\ a = -1, c = \sqrt{q}) \qquad\qquad (\text{A.29})$$

$$\frac{f(q,q^3)}{(q^2;q^2)_\infty} = \sum_{n=0}^{\infty} \frac{q^{n(2n+1)}}{(q)_{2n+1}} \qquad ([\text{Jac28, p. 179, line -3];}\ \textbf{S. 9}) \qquad (\text{A.30})$$

$$\frac{f(q,q^3)}{\psi(-q)} = \sum_{n=0}^{\infty} \frac{q^{n^2}(-1)_{2n}}{(q^2;q^2)_n(q^2;q^4)_n} \qquad\qquad (\textbf{S. 10c})\ (\text{A.31})$$

$$\frac{f(q,-q^3)}{\psi(-q)} = \sum_{n=0}^{\infty} \frac{q^{n^2}(-1;q^4)_n(-q;q^2)_n}{(q^2;q^2)_{2n}} \qquad\qquad (\textbf{S. 66})\ (\text{A.32})$$

$$\frac{f(-q,q^3)}{\psi(-q)} = \sum_{n=0}^{\infty} \frac{q^{n(n+2)}(-1;q^4)_n(-q;q^2)_n}{(q^2;q^2)_{2n}} \qquad\qquad (\textbf{S. 67})\ (\text{A.33})$$

$$\frac{f(q,q^3)}{\varphi(-q^2)} = \sum_{n=0}^{\infty} \frac{q^{n(n+1)}(-q)_{2n}}{(q;q^2)_{n+1}(q^4;q^4)_n} \qquad\qquad (\textbf{S. 11})\ (\text{A.34})$$

$$\frac{f(q,-q^3)}{\varphi(-q^2)} = \sum_{n=0}^{\infty} \frac{q^{n(n+1)}(-q^2;q^4)_n}{(q)_{2n+1}(-q;q^2)_n} \qquad\qquad (\textbf{S. 65})\ (\text{A.35})$$

$$\frac{f(-q^2,-q^2)}{\varphi(-q^2)} = \sum_{n=0}^{\infty} \frac{q^{n(n+1)}(q^2;q^2)_{n+1}}{(-q^3;q^3)_{n+1}(q)_n} \qquad ([\text{MSZ09, Eq. (2.3)]})\ (\text{A.36})$$

A.1.5 Mod 5 Identities

$$H(q) = \frac{f(-q,-q^4)}{(q)_\infty} = \sum_{n=0}^{\infty} \frac{q^{n(n+1)}}{(q)_n} \qquad ([\text{Rog94, p. 330 (2)];}\ \textbf{S. 14})$$

$$(\text{A.37})$$

$$= \sum_{n=0}^{\infty} \frac{q^{n^2}(q;q^2)_{n+1}}{(q)_n(q;q^2)_n} \qquad \text{([MSZ09, Eq. (2.5)])}$$

$$\text{(A.38)}$$

$$G(q) = \frac{f(-q^2,-q^3)}{(q)_\infty} = \sum_{n=0}^{\infty} \frac{q^{n^2}}{(q)_n} \qquad \text{([Rog94, p. 328 (2)]; \textbf{S. 18})}$$

$$\text{(A.39)}$$

$$= \sum_{n=0}^{\infty} \frac{q^{n(n+1)}(-q)_{n+1}}{(q^2;q^2)_n} \qquad \text{([MSZ09, Eq. (2.6)])}$$

$$\text{(A.40)}$$

$$\chi(-q)H(q) = \frac{f(-q,-q^4)}{(q^2;q^2)_\infty} = \sum_{n=0}^{\infty} \frac{(-1)^n q^{n(3n-2)}}{(-q;q^2)_n(q^4;q^4)_n}$$

$$\text{([Rog17, p. 330 (5)]; \textbf{S. 15})} \qquad \text{(A.41)}$$

$$= \sum_{n=0}^{\infty} \frac{(-1)^n q^{n(3n+2)}}{(-q;q^2)_{n+1}(q^4;q^4)_n}$$

$$\text{(Ramanujan [AB05, p. 252, Eq. (11.2.7)])}$$

$$\text{(A.42)}$$

$$\chi(-q)G(q) = \frac{f(-q^2,-q^3)}{(q^2;q^2)_\infty} = \sum_{n=0}^{\infty} \frac{(-1)^n q^{3n^2}}{(-q;q^2)_n(q^4;q^4)_n}$$

$$\text{([Rog94, p. 339, Ex. 2]; \textbf{S. 19})} \qquad \text{(A.43)}$$

$$\chi(-q^2)H(q) = \frac{f(-q,-q^4)}{\psi(-q)} = \sum_{n=0}^{\infty} \frac{q^{n(n+2)}}{(q^4;q^4)_n}$$

$$\text{([Rog17, p. 331, abv (7)]; \textbf{S. 16})} \qquad \text{(A.44)}$$

$$\chi(-q^2)G(q) = \frac{f(-q^2,-q^3)}{\psi(-q)} = \sum_{n=0}^{\infty} \frac{q^{n^2}}{(q^4;q^4)_n}$$

$$\text{([Rog94, p. 330]; \textbf{S. 20})} \qquad \text{(A.45)}$$

$$\frac{f(1,q^5)}{\psi(q)} = 2\sum_{n=0}^{\infty} \frac{(-1)^n q^{n(n+2)}(q;q^2)_n}{(-q;q^2)_{n+1}(q^4;q^4)_n}$$

$$\text{([MSZ09, Eq. (2.7)])} \qquad \text{(A.46)}$$

$$\frac{f(q,q^4)}{\psi(q)} = \sum_{n=0}^{\infty} \frac{(-1)^n q^{n(n+2)}(q;q^2)_n}{(-q;q^2)_n(q^4;q^4)_n}$$

$$\text{([BMS09, Eq. (2.17)])} \qquad \text{(A.47)}$$

$$\frac{f(q^2,q^3)}{\psi(q)} = \sum_{n=0}^{\infty} \frac{(-1)^n q^{n^2}(q;q^2)_n}{(-q;q^2)_n(q^4;q^4)_n} \qquad \textbf{(S. 21)} \quad \text{(A.48)}$$

$$\frac{H(q)\chi(-q)}{\chi(-q^2)} = \frac{f(-q,-q^4)}{\varphi(-q^2)} = \sum_{n=0}^{\infty} \frac{q^{n(n+1)}}{(q^2;q^2)_n(-q;q^2)_{n+1}} \qquad \textbf{(S. 17)} \quad \text{(A.49)}$$

$$\frac{G(q)\chi(-q)}{\chi(-q^2)} = \frac{f(-q^2,-q^3)}{\varphi(-q^2)} = \sum_{n=0}^{\infty} \frac{q^{n(n+1)}}{(q^2;q^2)_n(-q;q^2)_n} \qquad \textbf{(S. 99--)} \quad \text{(A.50)}$$

A.1.6 Mod 6 Identities

$$\frac{f(-q^3,-q^3)}{(q^2;q^2)_\infty} = \sum_{n=0}^{\infty} \frac{q^{2n^2}(q;q^2)_n}{(-q)_{2n}(q^2;q^2)_n}$$
$$([\text{BMS09, Eq. (2.13)}]) \qquad \text{(A.51)}$$

$$\frac{f(-q,-q^5)}{\psi(-q)} = \sum_{n=0}^{\infty} \frac{q^{n(n+2)}(-q;q^2)_n}{(q^4;q^4)_n}$$
$$(\text{Ramanujan [AB09, p. 87, Entry 4.2.11], Stanton [Sta01]})$$
$$\text{(A.52)}$$

$$\frac{f(-q^3,-q^3)}{\psi(-q)} = \sum_{n=0}^{\infty} \frac{q^{n^2}(-q;q^2)_n}{(q^4;q^4)_n}$$
$$(\text{Ramanujan [AB09, p. 85, Ent. 4.2.7]; } \textbf{S. 25}) \quad \text{(A.53)}$$

$$\frac{f(-q,-q^5)}{\psi(q)} = \sum_{n=0}^{\infty} \frac{(-1)^n q^{n^2}}{(q^2;q^2)_n} \quad (\text{sp. case of (1.100); } \textbf{S. 23}) \qquad \text{(A.54)}$$

$$\frac{f(-q,-q^5)}{\varphi(-q)} = \sum_{n=0}^{\infty} \frac{q^{n(n+1)}(-q)_n}{(q;q^2)_{n+1}(q)_n}$$
$$(\text{Ramanujan [AB09, p. 87, Entry 4.2.12],}$$
$$\text{Bailey [Bai35, p. 72, Eq. (10)]; } \textbf{S. 22}) \quad \text{(A.55)}$$

$$\frac{f(-q^3,-q^3)}{\varphi(-q)} = \sum_{n=0}^{\infty} \frac{q^{n^2}(-q)_n}{(q;q^2)_{n+1}(q)_n} \qquad \textbf{(S. 26)} \qquad \text{(A.56)}$$

$$= \sum_{n=0}^{\infty} \frac{q^{n^2}(-1)_n}{(q;q^2)_n(q)_n} \qquad ([\text{MSZ09}]) \qquad \text{(A.57)}$$

$$\frac{f(q,q^5)}{(q^2;q^2)_\infty} = \sum_{n=0}^\infty \frac{q^{2n(n+1)}(-q;q^2)_n}{(q)_{2n+1}(-q^2;q^2)_n} \tag{S. 27}$$
$$\tag{A.58}$$

$$= \sum_{n=0}^\infty \frac{q^{2n^2}(-q^{-1};q^2)_n(-q^3;q^2)_n}{(q^2;q^2)_{2n}}$$

(Ramanujan's (2.62) with $a=q$; Stanton [Sta01]) $\tag{A.59}$

$$= \sum_{n=0}^\infty \frac{q^{2n(n-1)}(-q;q^2)_n}{(q)_{2n}(-1;q^2)_{n+1}} \tag{A.60}$$

$$\frac{f(q^3,q^3)}{(q^2;q^2)_\infty} = \sum_{n=0}^\infty \frac{q^{2n^2}(-q;q^2)_n}{(q;q^2)_n(q^4;q^4)_n}$$

([BMS09, Eq. (2.13)]) $\tag{A.61}$

$$\frac{f(q,q^5)}{\varphi(-q^2)} = \sum_{n=0}^\infty \frac{q^{n(n+1)}(-q^2;q^2)_n}{(q)_{2n+1}}$$

(Ramanujan [AB05, p. 254, (11.3.5)]), (S. 28) $\tag{A.62}$

$$\frac{f(q^3,q^3)}{\varphi(-q^2)} = \sum_{n=0}^\infty \frac{q^{n(n+1)}(-1;q^2)_n}{(q)_{2n}} \quad \text{(S. 48–)} \tag{A.63}$$

$$\frac{f(q^2,q^4)}{\psi(-q)} = \sum_{n=0}^\infty \frac{q^{n^2}(-q;q^2)_n}{(q)_{2n}}$$

(Ramanujan [And81, p. 178 (3.1)], [AB09, p. 101, Entry 5.3.2]) $\tag{A.64}$

(S. 29)
$$\frac{1}{(q,q^4,q^5;q^6)_\infty} = \sum_{n=0}^\infty \frac{q^{n(3n-1)/2}(q^2;q^6)_n}{(q)_{3n}}$$

(special case of (1.103); cf. [CSS12, Eq. (1)]) $\tag{A.65}$

$$\frac{1}{(q,q^2,q^5;q^6)_\infty} = \sum_{n=0}^\infty \frac{q^{n(3n+1)/2}(q^4;q^6)_n}{(q)_{3n+1}}$$

(special case of (1.103); cf. [CSS12, Eq. (2)]) $\tag{A.66}$

A.1.7 Mod 7 Identities

$$\frac{f(-q,-q^6)}{(q^2;q^2)_\infty} = \sum_{n=0}^\infty \frac{q^{2n(n+1)}}{(q^2;q^2)_n(-q)_{2n+1}} \quad \text{([Rog17, p. 331 (6)]; S. 31)} \tag{A.67}$$

$$\frac{f(-q^2, -q^5)}{(q^2; q^2)_\infty} = \sum_{n=0}^{\infty} \frac{q^{2n(n+1)}}{(q^2; q^2)_n (-q)_{2n}} \qquad \text{([Rog94, p. 342]; S. 32)} \quad \text{(A.68)}$$

$$\frac{f(-q^3, -q^4)}{(q^2; q^2)_\infty} = \sum_{n=0}^{\infty} \frac{q^{2n^2}}{(q^2; q^2)_n (-q)_{2n}} \qquad \text{([Rog94, p. 339]; S. 33)} \quad \text{(A.69)}$$

A.1.8   Mod 8 Identities

A.1.8.1   Triple Products

$$\frac{f(-q, -q^7)}{\psi(-q)} = \sum_{n=0}^{\infty} \frac{q^{n(n+2)}(-q; q^2)_n}{(q^2; q^2)_n}$$

(Ramanujan [AB09, p. 37, Entry 1.7.12]; **S. 34**)   (A.70)

$$\frac{1}{(q^2, q^3, q^7; q^8)_\infty} = \sum_{n=0}^{\infty} \frac{q^{n(n+1)}(-q; q^2)_n}{(q^2; q^2)_n}$$

(Göllnitz [Göl67, (2.24)]; (1.102) with $a = -q^{1/2}$)   (A.71)

$$\frac{1}{(q, q^5, q^6; q^8)_\infty} = \sum_{n=0}^{\infty} \frac{q^{n(n+1)}(-q^{-1}; q^2)_n}{(q^2; q^2)_n}$$

(Göllnitz [Göl67, (2.22)]; (1.102) with $a = -q^{-1/2}$)   (A.72)

$$\frac{f(-q^3, -q^5)}{\psi(-q)} = \sum_{n=0}^{\infty} \frac{q^{n^2}(-q; q^2)_n}{(q^2; q^2)_n}$$

(Ramanujan [AB09, p. 36, Ent. 1.7.11; p. 88, Ent. 4.2.15];

S. 36)   (A.73)

$$\frac{f(-q, -q^7)}{\varphi(-q)} = \sum_{n=0}^{\infty} \frac{q^{n(n+3)/2}(-q; q^2)_n(-q)_n}{(q)_{2n+1}}$$

(Ramanujan [AB09, p. 32, Entry 1.7.6]; **S. 35**)   (A.74)

$$\frac{f(-q^2, -q^6)}{\varphi(-q)} = \sum_{n=0}^{\infty} \frac{q^{n(n+1)/2}(-q^2; q^2)_n}{(q; q^2)_{n+1}(q)_n}$$

(Ramanujan [AB09, p. 32, Entry 1.7.5])   (A.75)

$$\frac{f(-q^3, -q^5)}{\varphi(-q)} = \sum_{n=0}^{\infty} \frac{q^{n(n+1)/2}(-q; q^2)_n(-q)_n}{(q)_{2n+1}}$$

(Ramanujan [AB09, p. 34, Entry 1.7.8]; **S. 37**)   (A.76)

$$\frac{f(-q^4, -q^4)}{\varphi(-q)} = \sum_{n=0}^{\infty} \frac{q^{n(n+1)/2}(-1; q^2)_n}{(q)_n(q; q^2)_n}$$

(Ramanujan [AB09, p. 31, Entry 1.7.4])   (A.77)

$$\frac{f(q,q^7)}{(q^2;q^2)_\infty} = \sum_{n=0}^{\infty} \frac{q^{2n(n+1)}}{(q)_{2n+1}} \qquad \textbf{(S. 38)} \tag{A.78}$$

$$\frac{f(q^3,q^5)}{(q^2;q^2)_\infty} = \sum_{n=0}^{\infty} \frac{q^{2n^2}}{(q)_{2n}}$$

([Jac28, p. 170, 5th Eq.]; **S. 39**) $\tag{A.79}$

$$\frac{f(q,q^7)}{\psi(-q^2)} = \sum_{n=0}^{\infty} \frac{q^{2n^2}(-q^{-1};q^4)_n(-q^5;q^4)_n}{(q^8;q^8)_n(q^2;q^4)_n}$$

((2.63) with $a = q^{3/2}$) $\tag{A.80}$

$$\frac{f(q^3,q^5)}{\psi(-q^2)} = \sum_{n=0}^{\infty} \frac{q^{2n^2}(-q;q^2)_{2n}}{(q^8;q^8)_n(q^2;q^4)_n}$$

((2.63) $a = q^{\frac{1}{2}}$; Gessel–Stanton [GS83, p. 197, Eq. (7.24)]) $\tag{A.81}$

$$\frac{f(-q,-q^7)}{\varphi(-q^4)} = \sum_{n=0}^{\infty} \frac{q^{2n(n+1)}(-q^4;q^4)_n(q;q^2)_{2n+1}}{(q^4;q^4)_{2n+1}}$$

(sp. case (1.76); Gessel–Stanton [GS83, p. 197, Eq. (7.25)]) $\tag{A.82}$

$$\frac{f(-q^3,-q^5)}{\varphi(-q^4)} = \sum_{n=0}^{\infty} \frac{q^{2n(n+1)}(-q^4;q^4)_n(q^{-1};q^2)_{2n}}{(q^4;q^4)_{2n}}$$

((1.76) with $a = q^{-1/4}, b = q^{1/4}$) $\tag{A.83}$

A.1.8.2   Quintuple products

$$\frac{Q(q^4,q)}{(q)_\infty} = \sum_{n=0}^{\infty} \frac{q^{n^2}(-1;q^2)_n}{(q)_{2n}} \quad ((1.47) \text{ with } a = -1, c = \sqrt{q}; \textbf{S. 47}) \tag{A.84}$$

A.1.9   Mod 9 Identities

$$\frac{f(-q,-q^8)}{(q^3;q^3)_\infty} = \sum_{n=0}^{\infty} \frac{q^{3n(n+1)}(q)_{3n+1}}{(q^3;q^3)_n(q^3;q^3)_{2n+1}} \tag{A.85}$$

(Bailey [Bai47, p. 422, Eq. (1.7)]; **S. 40c**)

$$\frac{f(-q^2,-q^7)}{(q^3;q^3)_\infty} = \sum_{n=0}^{\infty} \frac{q^{3n(n+1)}(q)_{3n}(1-q^{3n+2})}{(q^3;q^3)_n(q^3;q^3)_{2n+1}} \tag{A.86}$$

(Bailey [Bai47, p. 422, Eq. (1.8)]; **S. 41c**)

$$\frac{f(-q^4, -q^5)}{(q^3; q^3)_\infty} = \sum_{n=0}^{\infty} \frac{q^{3n^2}(q)_{3n}}{(q^3; q^3)_n (q^3; q^3)_{2n}} \tag{A.87}$$

$$((2.62)\ a = -q^{\frac{1}{2}};\ \text{Bailey [Bai47, p. 422, Eq. (1.6)]};\ \textbf{S. 42c})$$

$$\frac{f(-q^3, -q^6)}{\psi(-q)} = \sum_{n=0}^{\infty} \frac{q^{n^2}(-1; q^6)_n(-q; q^2)_n}{(-1; q^2)_n (q^2; q^2)_{2n}} \tag{A.88}$$

$$([\text{MS08, p. 767, Eq. (1.13)}])$$

$$\frac{f(-q^3, q^6)}{\psi(-q)} = 1 + \sum_{n=1}^{\infty} \frac{q^{n^2}(q^6; q^6)_{n-1}(-q; q^2)_n}{(q^2; q^2)_{2n}(q^2; q^2)_{n-1}} \qquad (\textbf{S. 113}) \tag{A.89}$$

A.1.10   Mod 10 Identities

A.1.10.1   Triple Products

$$\frac{f(-q, -q^9)}{\varphi(-q)} = \sum_{n=0}^{\infty} \frac{q^{n(n+3)/2}(-q)_n}{(q; q^2)_{n+1}(q)_n}$$

$$([\text{Rog17, p. 330 (4), line 2}];\ \textbf{S. 43}) \tag{A.90}$$

$$\frac{f(-q^3, -q^7)}{\varphi(-q)} = \sum_{n=0}^{\infty} \frac{q^{n(n+1)/2}(-q)_n}{(q; q^2)_{n+1}(q)_n}$$

$$([\text{Rog17, p. 330 (4), line 1}];\ \textbf{S. 45}) \tag{A.91}$$

$$\frac{f(-q^5, -q^5)}{\varphi(-q)} = \sum_{n=0}^{\infty} \frac{q^{n(n+1)/2}(-1)_n}{(q; q^2)_n(q)_n}$$

$$([\text{Rog17, p. 330 (4), line 3, corrected}]) \tag{A.92}$$

$$\frac{H(q^2)}{\chi(-q)} = \frac{f(-q^2, -q^8)}{(q)_\infty} = \sum_{n=0}^{\infty} \frac{q^{3n(n+1)/2}}{(q; q^2)_{n+1}(q)_n}$$

$$([\text{Rog17, p. 330 (2), line 2}];\ \textbf{S. 44}) \tag{A.93}$$

$$\frac{G(q^2)}{\chi(-q)} = \frac{f(-q^4, -q^6)}{(q)_\infty} = \sum_{n=0}^{\infty} \frac{q^{n(3n-1)/2}}{(q; q^2)_n(q)_n}$$

$$([\text{Rog94, p. 341, Ex. 1}];\ \textbf{S. 46}) \tag{A.94}$$

$$= \sum_{n=0}^{\infty} \frac{q^{n(3n+1)/2}}{(q; q^2)_{n+1}(q)_n}$$

$$([\text{Rog17, p. 330 (2), line 1}]) \tag{A.95}$$

A.1.10.2 Quintuple Products

$$\frac{G(q^2)}{\chi(-q)} = \frac{Q(q^5, -q)}{\varphi(-q)} = \sum_{n=0}^{\infty} \frac{q^{n(3n+1)/2}(-q)_n}{(q)_{2n+1}} \qquad \textbf{(S. 62)}$$

$$\text{(A.96)}$$

$$= \sum_{n=0}^{\infty} \frac{q^{n(3n-1)/2}(-q)_n}{(q)_{2n}}$$

$$\text{([Rog17, p. 332 (12), 1st Eq.])} \qquad \text{(A.97)}$$

$$\frac{H(q^2)}{\chi(-q)} = \frac{Q(q^5, -q^2)}{\varphi(-q)} = \sum_{n=0}^{\infty} \frac{q^{3n(n+1)/2}(-q)_n}{(q)_{2n+1}}$$

$$\text{([Rog17, p. 332 (12), 2nd Eq.]; } \textbf{S. 63}\text{)} \qquad \text{(A.98)}$$

A.1.12 Mod 12 Identities

A.1.12.1 Triple Products

$$\frac{f(-q, -q^{11})}{(q)_\infty} = \sum_{n=0}^{\infty} \frac{q^{n(n+2)}(-q^2; q^2)_n(1 - q^{n+1})}{(q)_{2n+2}} \qquad \textbf{(S. 49c)} \qquad \text{(A.99)}$$

$$\frac{f(-q^2, -q^{10})}{(q)_\infty} = \sum_{n=0}^{\infty} \frac{q^{n(n+2)}(-q; q^2)_n}{(q)_{2n+1}}$$

$$\text{(Ramanujan [AB09, p. 65, Entry 3.4.4]; } \textbf{S. 50}\text{)} \qquad \text{(A.100)}$$

$$\frac{f(-q^3, -q^9)}{(q)_\infty} = \sum_{n=0}^{\infty} \frac{q^{n(n+1)}(-q^2; q^2)_n}{(q)_{2n+1}} \qquad \textbf{(S. 28)} \qquad \text{(A.101)}$$

$$\frac{f(-q^4, -q^8)}{(q)_\infty} = \sum_{n=0}^{\infty} \frac{q^{n(n+1)}(-q; q^2)_n}{(q)_{2n+1}}$$

$$\text{((1.47) with } aq = -c = q^{3/2}; \textbf{S. 51}\text{)} \qquad \text{(A.102)}$$

$$\frac{f(-q^5, -q^7)}{(q)_\infty} = \sum_{n=0}^{\infty} \frac{q^{n^2}(-q^2; q^2)_{n-1}(1 + q^n)}{(q)_{2n}} \qquad \textbf{(S. 54c)} \qquad \text{(A.103)}$$

$$\frac{f(-q^6, -q^6)}{(q)_\infty} = \sum_{n=0}^{\infty} \frac{q^{n^2}(-q; q^2)_n}{(q)_{2n}}$$

$$\text{(Ramanujan [AB05, Ent. 11.3.1]; } \textbf{S. 29}\text{)} \qquad \text{(A.104)}$$

$$\frac{f(-q, -q^{11})}{(q^4; q^4)_\infty} = \sum_{n=0}^{\infty} \frac{q^{4n(n+1)}(q; q^2)_{2n+1}}{(q^4; q^4)_{2n+1}} \qquad \textbf{(S. 55)} \qquad \text{(A.105)}$$

$$\frac{f(-q^5, -q^7)}{(q^4; q^4)_\infty} = \sum_{n=0}^{\infty} \frac{q^{4n^2}(q; q^2)_{2n}}{(q^4; q^4)_{2n}}$$

$$((2.62) \text{ with } a = -\sqrt{q}; \textbf{ S. 53}) \tag{A.106}$$

$$\frac{f(-q^3, -q^9)}{\psi(-q)} = \sum_{n=0}^{\infty} \frac{q^{2n(n+1)}(-q; q^2)_n}{(q; q^2)_{n+1}(q^4; q^4)_n} \qquad (\textbf{S. 27}) \tag{A.107}$$

$$\frac{f(-q^4, -q^8)}{\psi(-q)} = \sum_{n=0}^{\infty} \frac{q^{n(2n-1)}(-q; q^2)_n}{(q^2; q^2)_n(q^2; q^4)_n} \qquad (\textbf{S. 52}) \tag{A.108}$$

$$\frac{f(-q^5, -q^7)}{\psi(-q^3)} = \sum_{n=0}^{\infty} \frac{q^{3n^2}(q^2; q^2)_{3n}}{(q^{12}; q^{12})_n(q^3; q^3)_{2n}}$$

$$((2.63) \ a = -q^{1/3}; \text{ Dyson [Bai48, p. 9, Eq. (7.5)]}) \tag{A.109}$$

$$\frac{f(-q, -q^{11})}{\psi(-q^3)} = \sum_{n=0}^{\infty} \frac{q^{3n^2}(q^2; q^2)_{3n+1}}{(q^{12}; q^{12})_n(q^3; q^3)_{2n}(q^{6n} - q^2)}$$

$$(\text{Dyson [Bai48, p. 9, Eq. (7.6)]}) \tag{A.110}$$

$$\frac{f(q, q^{11})}{(q)_\infty} = \sum_{n=0}^{\infty} \frac{q^{n(n+2)}(-q)_n}{(q; q^2)_{n+1}(q)_{n+1}} \qquad (\textbf{S. 56}) \tag{A.111}$$

$$\frac{f(q^3, q^9)}{(q)_\infty} = \frac{1 + q^3}{(1-q)(1-q^2)} + \sum_{n=1}^{\infty} \frac{q^{n(n+2)}(-q)_{n-1}(-q)_{n+2}}{(q)_{2n+2}} \quad \text{[MSZ09, (2.10)]}$$

$$\tag{A.112}$$

$$\frac{f(q^5, q^7)}{(q)_\infty} = 1 + \sum_{n=1}^{\infty} \frac{q^{n^2}(-q)_{n-1}}{(q; q^2)_n(q)_n} \qquad (\textbf{S. 58c}) \tag{A.113}$$

$$\frac{f(q, q^{11})}{(q^4; q^4)_\infty} = \sum_{n=0}^{\infty} \frac{q^{4n(n+1)}(-q; q^2)_{2n+1}}{(q^4; q^4)_{2n+1}} \qquad (\textbf{S. 57}) \tag{A.114}$$

$$\frac{f(q^3, q^9)}{(q^4; q^4)_\infty} = \sum_{n=0}^{\infty} \frac{q^{n(n+2)}(-q; q^2)_n}{(q^4; q^4)_n} \quad (\text{Ramanujan [AB09, p. 105, Entry 5.3.7]})$$

$$\tag{A.115}$$

$$\frac{f(q^5, q^7)}{(q^4; q^4)_\infty} = \sum_{n=0}^{\infty} \frac{q^{4n^2}(-q; q^2)_{2n}}{(q^4; q^4)_{2n}} \quad ((2.62) \text{ with } a = \sqrt{q}; \textbf{ S. 53}-) \tag{A.116}$$

A.1.12.2 Quintuple Products

$$\frac{Q(q^6, -q)}{\varphi(-q)} = \sum_{n=0}^{\infty} \frac{q^{n(n+1)/2}(-1; q^3)_n(-q)_n}{(q)_{2n}(-1)_n} \tag{A.117}$$

([MS08, p. 767, Eq. (1.22)])

$$\frac{Q(q^6, -q^2)}{\varphi(-q)} = \sum_{n=0}^{\infty} \frac{q^{n(n+1)/2}(-q^3; q^3)_n}{(q)_{2n+1}} \tag{A.118}$$

([MS08, p. 768, Eq. (1.24)])

$$\frac{Q(q^6, q)}{\varphi(-q)} = 1 + \sum_{n=1}^{\infty} \frac{q^{n(n+1)/2}(q^3; q^3)_{n-1}(-q)_n}{(q)_{2n}(q)_{n-1}} \tag{A.119}$$

([MS08, p. 768, Eq. (1.27)])

$$\frac{Q(q^6, q^2)}{\varphi(-q)} = \sum_{n=0}^{\infty} \frac{q^{n(n+1)/2}(q^3; q^3)_n(-q)_n}{(q)_{2n+1}(q)_n} \tag{A.120}$$

(Dyson [MS08, p. 434, Eq. (D2)])

A.1.14 Mod 14 Identities

A.1.14.1 Triple Products

$$\frac{f(-q^2, -q^{12})}{(q)_\infty} = \sum_{n=0}^{\infty} \frac{q^{n(n+2)}}{(q; q^2)_{n+1}(q)_n} \qquad \text{([Rog17, p. 329 (1)]; \textbf{S. 59})} \tag{A.121}$$

$$\frac{f(-q^4, -q^{10})}{(q)_\infty} = \sum_{n=0}^{\infty} \frac{q^{n(n+1)}}{(q; q^2)_{n+1}(q)_n} \qquad \text{([Rog17, p. 329 (1)]; \textbf{S. 60})} \tag{A.122}$$

$$\frac{f(-q^6, -q^8)}{(q)_\infty} = \sum_{n=0}^{\infty} \frac{q^{n^2}}{(q; q^2)_n(q)_n} \qquad \text{([Rog94, p. 341, Ex. 2]; \textbf{S. 61})} \tag{A.123}$$

A.1.14.2 Quintuple Products

$$\frac{Q(q^7, -q)}{\varphi(-q)} = \sum_{n=0}^{\infty} \frac{q^{n(n+1)/2}(-q)_n}{(q)_{2n}} \qquad \textbf{(S. 81)} \tag{A.124}$$

$$\frac{Q(q^7, -q^2)}{\varphi(-q)} = \sum_{n=0}^{\infty} \frac{q^{n(n+1)/2}(-q)_n}{(q)_{2n+1}} \qquad \textbf{(S. 80)} \tag{A.125}$$

$$\frac{Q(q^7, -q^3)}{\varphi(-q)} = \sum_{n=0}^{\infty} \frac{q^{n(n+3)/2}(-q)_n}{(q)_{2n+1}} \qquad \textbf{(S. 82)} \tag{A.126}$$

A.1.15 Mod 15 Identities

$$\frac{f(-q,-q^{14})}{(q^5;q^5)_\infty} = 1 - \sum_{n=1}^{\infty} \frac{q^{5n^2-4}(q;q^5)_{n-1}(q^4;q^5)_{n+1}}{(q^5;q^5)_{2n}} \quad ((2.62)\ a = -q^{13/5})$$
$$(A.127)$$

$$\frac{f(-q^2,-q^{13})}{(q^5;q^5)_\infty} = 1 - \sum_{n=1}^{\infty} \frac{q^{5n^2-3}(q^2;q^5)_{n-1}(q^3;q^5)_{n+1}}{(q^5;q^5)_{2n}} \quad ((2.62)\ a = -q^{11/5})$$
$$(A.128)$$

$$\frac{f(-q^3,-q^{12})}{(q^5;q^5)_\infty} = 1 - \sum_{n=1}^{\infty} \frac{q^{5n^2-2}(q^3;q^5)_{n-1}(q^2;q^5)_{n+1}}{(q^5;q^5)_{2n}} \quad ((2.62)\ a = -q^{9/5})$$
$$(A.129)$$

$$\frac{f(-q^4,-q^{11})}{(q^5;q^5)_\infty} = 1 - \sum_{n=1}^{\infty} \frac{q^{5n^2-1}(q;q^5)_{n-1}(q;q^5)_{n+1}}{(q^5;q^5)_{2n}} \quad ((2.62)\ a = -q^{7/5})$$
$$(A.130)$$

$$\frac{f(-q^6,-q^9)}{(q^5;q^5)_\infty} = \sum_{n=0}^{\infty} \frac{q^{5n^2}(q;q^5)_n(q^4;q^5)_n}{(q^5;q^5)_{2n}} \quad ((2.62)\ a = -q^{3/5}) \qquad (A.131)$$

$$\frac{f(-q^7,-q^8)}{(q^5;q^5)_\infty} = \sum_{n=0}^{\infty} \frac{q^{5n^2}(q^2;q^5)_n(q^3;q^5)_n}{(q^5;q^5)_{2n}} \quad ((2.62)\ \text{with}\ a = -q^{1/5})$$
$$(A.132)$$

A.1.16 Mod 16 Identities

A.1.16.1 Triple Products

$$\frac{f(-q^2,-q^{14})}{\psi(-q)} = \sum_{n=0}^{\infty} \frac{q^{n(n+2)}(-q;q^2)_n(-q^4;q^4)_n}{(-q^2;q^2)_{n+1}(q^2;q^4)_{n+1}(q^2;q^2)_n} \qquad (\mathbf{S.\ 68}) \quad (A.133)$$

$$\frac{f(-q^4,-q^{12})}{\psi(-q)} = \sum_{n=0}^{\infty} \frac{q^{n(n+2)}(-q^2;q^4)_n}{(q;q^2)_{n+1}(q^4;q^4)_n} \qquad (\mathbf{S.\ 70}) \quad (A.134)$$

$$\frac{f(-q^6,-q^{10})}{\psi(-q)} = \sum_{n=0}^{\infty} \frac{q^{n^2}(-q^4;q^4)_{n-1}}{(q)_{2n}(-q^2;q^2)_{n-1}} \qquad (\mathbf{S.\ 71}) \quad (A.135)$$

$$\frac{f(-q^8,-q^8)}{\psi(-q)} = \sum_{n=0}^{\infty} \frac{q^{n^2}(-q^2;q^4)_n}{(q;q^2)_n(q^4;q^4)_n}$$
$$([\text{Sil07, Eq. (5.5)}];\ (1.77)\ b = ic = iq^{1/2}) \quad (A.136)$$

$$\frac{f(q^2, q^{14})}{\psi(-q)} = \sum_{n=0}^{\infty} \frac{(-q^2; q^2)_n q^{n(n+2)}}{(q)_{2n+2}} \qquad \textbf{(S. 69)} \tag{A.137}$$

$$= \sum_{n=0}^{\infty} \frac{q^{n(n+1)/2+n}(-q)_n}{(q)_{n+1}} \qquad \text{(Gessel–Stanton [GS83, Eq. (7.15)])} \tag{A.138}$$

$$= 1 + \frac{q}{(1-q)(1-q^2)} + \sum_{n=2}^{\infty} \frac{q^{n^2-2}(-q^2; q^2)_{n-2}(1 + q^{2n+2})}{(q)_{2n}} \tag{A.139}$$

([MSZ09])

$$\frac{f(q^6, q^{10})}{\psi(-q)} = 1 + \sum_{n=1}^{\infty} \frac{q^{n^2}(-q)_{2n-1}}{(q^2; q^4)_n (q^2; q^2)_n} \qquad \textbf{(S. 72)} \tag{A.140}$$

$$= 1 + \sum_{n=1}^{\infty} \frac{q^{n(n+1)/2}(-q)_{n-1}}{(q)_n} \qquad \text{(Gessel–Stanton [GS83, Eq. (7.13)])} \tag{A.141}$$

A.1.16.2 Quintuple Products

$$\frac{Q(q^8, -q)}{(q)_\infty} = \sum_{n=0}^{\infty} \frac{q^{2n^2}}{(q)_{2n}} \qquad \textbf{(S. 83)} \tag{A.142}$$

$$\frac{Q(q^8, -q^2)}{(q)_\infty} = \sum_{n=0}^{\infty} \frac{q^{n(2n+1)}}{(q)_{2n+1}} \qquad \textbf{(S. 84)} \tag{A.143}$$

$$= \sum_{n=0}^{\infty} \frac{q^{n(2n-1)}}{(q)_{2n}} \qquad \text{([Sta31, p. 809, Eq. (3.29)]; } \textbf{S. 85)} \tag{A.144}$$

$$\frac{Q(q^8, -q^3)}{(q)_\infty} = \sum_{n=0}^{\infty} \frac{q^{2n(n+1)}}{(q)_{2n+1}} \qquad \textbf{(S. 86)} \tag{A.145}$$

A.1.18 Mod 18 Identities
A.1.18.1 Triple Products

$$\frac{f(-q^3, -q^{15})}{\varphi(-q)} = \sum_{n=0}^{\infty} \frac{q^{n(n+3)/2}(q^3; q^3)_n(-q)_{n+1}}{(q)_{2n+2}(q)_n}$$
$$\text{(Dyson [Bai47, p. 434 (D1)]; } \textbf{S. 76)} \tag{A.146}$$

$$\frac{f(-q^6, -q^{12})}{\varphi(-q)} = \sum_{n=0}^{\infty} \frac{q^{n(n+1)/2}(-q)_n(q^3; q^3)_n}{(q)_n(q)_{2n+1}}$$

$$\text{(Dyson [Bai47, p. 434 (D2)]; \textbf{S. 77c})} \tag{A.147}$$

$$\frac{f(-q^9, -q^9)}{\varphi(-q)} = 1 + \sum_{n=1}^{\infty} \frac{q^{n(n+1)/2}(q^3; q^3)_{n-1}(-1)_{n+1}}{(q)_{2n}(q)_{n-1}}$$

$$\text{(Dyson [Bai47, p. 434 (D3)]; \textbf{S. 78})} \tag{A.148}$$

A.1.18.2 Quintuple Products

$$\frac{Q(q^9, -q)}{(q)_\infty} = \sum_{n=0}^{\infty} \frac{q^{n(n+1)}(-1; q^3)_n}{(-1)_n(q)_{2n}} \quad \text{([Lox84, p. 158, Eq. (P12)]} \tag{A.149}$$

$$\frac{Q(q^9, -q^2)}{(q)_\infty} = \sum_{n=0}^{\infty} \frac{q^{n^2}(-1; q^3)_n}{(-1)_n(q)_{2n}} \quad \text{([Lox84, p. 159, (P12 bis))]} \tag{A.150}$$

$$\frac{Q(q^9, -q^3)}{(q)_\infty} = \sum_{n=0}^{\infty} \frac{q^{n(n+1)}(-q^3; q^3)_n}{(-q)_n(q)_{2n+1}} \quad \text{([MS08, p. 766, Eq. (1.5)])} \tag{A.151}$$

$$\frac{Q(q^9, -q^4)}{(q)_\infty} = \sum_{n=0}^{\infty} \frac{q^{n(n+2)}(-q^3; q^3)_n(1 - q^{n+1})}{(-q)_n(q)_{2n+2}} \quad \text{([MS08, p. 767, Eq. (1.6)])}$$
$$\tag{A.152}$$

$$\frac{Q(q^9, q)}{(q)_\infty} = 1 + \sum_{n=1}^{\infty} \frac{q^{n^2}(q^3; q^3)_{n-1}(2 + q^n)}{(q)_{n-1}(q)_{2n}} \quad \text{([MS08, p. 766, Eq. (1.7)])}$$
$$\tag{A.153}$$

$$\frac{Q(q^9, q^2)}{(q)_\infty} = 1 + \sum_{n=1}^{\infty} \frac{q^{n^2}(q^3; q^3)_{n-1}(1 + 2q^n)}{(q)_{n-1}(q)_{2n}} \quad \text{([MS08, p. 766, Eq. (1.8)])}$$
$$\tag{A.154}$$

$$\frac{Q(q^9, q^3)}{(q)_\infty} = \sum_{n=0}^{\infty} \frac{q^{n(n+1)}(q^3; q^3)_n}{(q)_n(q)_{2n+1}} \quad \text{(Dyson [Bai47, p. 433, Eq. (B3)])}$$
$$\tag{A.155}$$

$$\frac{Q(q^9, q^4)}{(q)_\infty} = \sum_{n=0}^{\infty} \frac{q^{n(n+2)}(q^3; q^3)_n}{(q)_n^2(q^{n+2})_{n+1}} \quad \text{([MS08, p. 766, Eq. (1.10)])} \tag{A.156}$$

A.1.20 Mod 20 Identities
A.1.20.1 Triple Products

$$\frac{H(q^4)}{\chi(-q)} = \frac{f(-q^4, -q^{16})}{\psi(-q)} = \sum_{n=0}^{\infty} \frac{q^{n(n+2)}}{(q)_{2n+1}} \quad \text{([Rog17, p. 330 (3), 2nd Eq.])}$$
$$\tag{A.157}$$

$$\frac{G(q^4)}{\chi(-q)} = \frac{f(-q^8, -q^{12})}{\psi(-q)} = \sum_{n=0}^{\infty} \frac{q^{n^2}}{(q)_{2n}} \qquad \text{([Rog94, p. 330 (3), 1st Eq.]; \textbf{S. 79})}$$

$$\text{(A.158)}$$

A.1.20.2 Quintuple Products

$$\frac{G(-q)}{\chi(-q)} = \frac{Q(q^{10}, -q)}{(q)_{\infty}} = \sum_{n=0}^{\infty} \frac{q^{n(n+1)}}{(q)_{2n}}$$

$$\text{([Rog94, p. 332, Eq. (13)]; \textbf{S. 99})} \qquad \text{(A.159)}$$

$$\frac{G(q^4)}{\chi(-q)} = \frac{Q(q^{10}, -q^2)}{(q)_{\infty}} = \sum_{n=0}^{\infty} \frac{q^{n^2}}{(q)_{2n}}$$

$$\text{([Rog94, p. 331 above (5)]; \textbf{S. 98})} \qquad \text{(A.160)}$$

$$\frac{H(-q)}{\chi(-q)} = \frac{Q(q^{10}, -q^3)}{(q)_{\infty}} = \sum_{n=0}^{\infty} \frac{q^{n(n+1)}}{(q)_{2n+1}}$$

$$\text{([Rog94, p. 331, Eq. (6)]; \textbf{S. 94})} \qquad \text{(A.161)}$$

$$\frac{H(q^4)}{\chi(-q)} = \frac{Q(q^{10}, -q^4)}{(q)_{\infty}} = \sum_{n=0}^{\infty} \frac{q^{n(n+2)}}{(q)_{2n+1}}$$

$$\text{([Rog94, p. 331, Eq. (7)]; \textbf{S. 96})} \qquad \text{(A.162)}$$

$$\frac{Q(q^{10}, -q)}{\psi(-q)} = \sum_{n=0}^{\infty} \frac{q^{3n^2}(-q; q^2)_n}{(q^2; q^2)_{2n}} \qquad \text{(\textbf{S. 100c})}$$

$$\text{(A.163)}$$

$$\frac{Q(q^{10}, -q^3)}{\psi(-q)} = \sum_{n=0}^{\infty} \frac{q^{n(3n-2)}(-q; q^2)_n}{(q^2; q^2)_{2n}} \qquad \text{(\textbf{S. 95})}$$

$$\text{(A.164)}$$

$$= \sum_{n=0}^{\infty} \frac{q^{n(3n+2)}(-q; q^2)_{n+1}}{(q^2; q^2)_{2n+1}} \qquad \text{(\textbf{S. 97c})}$$

$$\text{(A.165)}$$

A.1.24 Mod 24 Identities

$$\frac{Q(q^{12}, -q)}{\psi(-q)} = \sum_{n=0}^{\infty} \frac{q^{n(n+2)}(-q; q^2)_n(-1; q^6)_n}{(q^2; q^2)_{2n}(-1; q^2)_n}$$

$$\text{([MS08, p. 767, Eq. (1.12)])} \qquad \text{(A.166)}$$

$$\frac{Q(q^{12}, -q^2)}{\psi(-q)} = \sum_{n=0}^{\infty} \frac{q^{n^2}(-q^3; q^6)_n}{(q^2; q^2)_{2n}}$$

$$\text{(Ramanujan [AB09, p. 105, Entry 5.3.8])} \qquad (A.167)$$

$$\frac{Q(q^{12}, -q^3)}{\psi(-q)} = \sum_{n=0}^{\infty} \frac{q^{n^2}(-q; q^2)_n(-1; q^6)_n}{(-1; q^2)_n(q^2; q^2)_{2n}}$$

$$\text{([MS08, p. 767, Eq. (1.13)])} \qquad (A.168)$$

$$\frac{Q(q^{12}, -q^4)}{\psi(-q)} = \sum_{n=0}^{\infty} \frac{q^{n(n+2)}(-q^3; q^6)_n}{(q^2; q^2)_{2n}(1 - q^{2n+1})}$$

$$\text{([MS08, p. 767, Eq. (1.14)])} \qquad (A.169)$$

$$\frac{Q(q^{12}, -q^5)}{\psi(-q)} = \sum_{n=0}^{\infty} \frac{q^{n(n+2)}(-q; q^2)_{n+1}(-q^6; q^6)_n(1 - q^{2n+2})}{(-q^2; q^2)_n(q^2; q^2)_{2n+2}}$$

$$\text{([MS08, p. 767, Eq. (1.15)])} \qquad (A.170)$$

$$\frac{Q(q^{12}, q)}{\psi(-q)} = 1 + \sum_{n=1}^{\infty} \frac{q^{n^2}(-q; q^2)_n(q^6; q^6)_{n-1}(2 + q^{2n})}{(q^2; q^2)_{2n}(q^2; q^2)_{n-1}}$$

$$\text{([MS08, p. 767, Eq. (1.17)])} \qquad (A.171)$$

$$\frac{Q(q^{12}, q^2)}{\psi(-q)} = \sum_{n=0}^{\infty} \frac{q^{n^2}(q^3; q^6)_n}{(q; q^2)_n^2(q^4; q^4)_n}$$

$$\text{(Ramanujan [AB09, p. 105, Entry 5.3.9])} \qquad (A.172)$$

$$\frac{Q(q^{12}, q^3)}{\psi(-q)} = 1 + \sum_{n=1}^{\infty} \frac{q^{n^2}(-q; q^2)_n(q^6; q^6)_{n-1}(1 + 2q^{2n})}{(q^2; q^2)_{2n}(q^2; q^2)_{n-1}}$$

$$\text{([MS08, p. 767, Eq. (1.18)])} \qquad (A.173)$$

$$\frac{Q(q^{12}, q^4)}{\psi(-q)} = \sum_{n=0}^{\infty} \frac{q^{n(n+2)}(q^3; q^6)_n(-q; q^2)_{n+1}}{(q^2; q^2)_{2n+1}(q; q^2)_n} \qquad \textbf{(S. 110c)}$$

$$(A.174)$$

$$\frac{Q(q^{12}, q^5)}{\psi(-q)} = \sum_{n=0}^{\infty} \frac{q^{n(n+2)}(q^6; q^6)_n(-q; q^2)_{n+1}(1 - q^{2n+2})}{(q^2; q^2)_{2n+2}(q^2; q^2)_n} \qquad \textbf{(S. 108c)}$$

$$(A.175)$$

$$\frac{Q(q^{12}, -q)}{\varphi(-q^2)} = \sum_{n=0}^{\infty} \frac{q^{n(n+1)}(-q^2; q^2)_n(-q^3; q^6)_n}{(q)_{2n}(-q)_{2n+1}(-q; q^2)_n}$$

$$\text{([MS08, p. 767, Eq. (1.21)])} \qquad (A.176)$$

$$\frac{Q(q^{12}, -q^3)}{\varphi(-q^2)} = \sum_{n=0}^{\infty} \frac{q^{n(n+1)}(-q^2; q^2)_n(-q^3; q^6)_n}{(q^2; q^2)_{2n+1}(-q; q^2)_n}$$

$$\text{([MS08, p. 767, Eq. (1.23)])} \qquad (A.177)$$

$$\frac{Q(q^{12},-q^5)}{\varphi(-q^2)} = \sum_{n=0}^{\infty} \frac{q^{n(n+3)}(-q^2;q^2)_n(-q^3;q^6)_n}{(q^2;q^2)_{2n+1}(-q;q^2)_n}$$

([MS08, p. 768, Eq. (1.25)]) \hfill (A.178)

$$\frac{Q(q^{12},q)}{\varphi(-q^2)} = \sum_{n=0}^{\infty} \frac{q^{n(n+1)}(-q^2;q^2)_n(q^3;q^6)_n}{(q)_{2n+1}(-q)_{2n}(q;q^2)_n}$$

([MS08, p. 768, Eq. (1.26)]) \hfill (A.179)

$$\frac{Q(q^{12},q^3)}{\varphi(-q^2)} = \sum_{n=0}^{\infty} \frac{q^{n(n+1)}(q^3;q^6)_n(-q^2;q^2)_n}{(q^2;q^2)_{2n+1}(q;q^2)_n} \qquad \textbf{(S. 107)} \qquad \text{(A.180)}$$

$$\frac{Q(q^{12},q^5)}{\varphi(-q^2)} = \sum_{n=0}^{\infty} \frac{q^{n(n+3)}(q^3;q^6)_n(-q^2;q^2)_n}{(q^2;q^2)_{2n+1}(q;q^2)_n}$$

([MS08, p. 768, Eq. (1.30)]) \hfill (A.181)

A.1.27   Dyson's Mod 27 Identities

$$\frac{f(-q^3,-q^{24})}{(q)_\infty} = \sum_{n=0}^{\infty} \frac{q^{n(n+3)}(q^3;q^3)_n}{(q)_{2n+2}(q)_n} \qquad \text{(Dyson [Bai47, p. 434 (B1)]; \textbf{S. 90})}$$
\hfill (A.182)

$$\frac{f(-q^6,-q^{21})}{(q)_\infty} = \sum_{n=0}^{\infty} \frac{q^{n(n+2)}(q^3;q^3)_n}{(q)_{2n+2}(q)_n} \qquad \text{(Dyson [Bai47, p. 434 (B2)]; \textbf{S. 91})}$$
\hfill (A.183)

$$\frac{f(-q^9,-q^{18})}{(q)_\infty} = \sum_{n=0}^{\infty} \frac{q^{n(n+1)}(q^3;q^3)_n}{(q)_{2n+1}(q)_n} \qquad \text{(Dyson [Bai47, p. 434 (B3)]; \textbf{S. 92})}$$
\hfill (A.184)

$$\frac{f(-q^{12},-q^{15})}{(q)_\infty} = 1 + \sum_{n=1}^{\infty} \frac{q^{n^2}(q^3;q^3)_{n-1}}{(q)_{2n-1}(q)_n} \qquad \text{(Dyson [Bai47, p. 434 (B4)]; \textbf{S. 93})}$$
\hfill (A.185)

A.1.28   Mod 28 Identities

$$\frac{Q(q^{14},-q)}{\psi(-q)} = \sum_{n=0}^{\infty} \frac{q^{n(n+2)}(-q;q^2)_n}{(q^2;q^2)_{2n}} \qquad \textbf{(S. 118)} \qquad \text{(A.186)}$$

$$\frac{Q(q^{14},-q^3)}{\psi(-q)} = \sum_{n=0}^{\infty} \frac{q^{n^2}(-q;q^2)_n}{(q^2;q^2)_{2n}} \qquad \textbf{(S. 117)} \qquad \text{(A.187)}$$

$$\frac{Q(q^{14}, -q^5)}{\psi(-q)} = \sum_{n=0}^{\infty} \frac{q^{n(n+2)}(-q; q^2)_{n+1}}{(q^2; q^2)_{2n+1}} \qquad \textbf{(S. 119)} \qquad (A.188)$$

A.1.32 Mod 32 Identities

$$\frac{Q(q^{16}, -q^2)}{(q)_\infty} = 1 + \sum_{n=1}^{\infty} \frac{q^{n^2}(-q^2; q^2)_{n-1}}{(q)_{2n}} \qquad \textbf{(S. 121)}$$
$$(A.189)$$

$$\frac{Q(q^{16}, -q^4)}{(q)_\infty} = \sum_{n=0}^{\infty} \frac{q^{n(n+1)}(-q; q^2)_n}{(q)_{2n+1}} \qquad ((1.47) \text{ with } aq = -c = q^{3/2})$$
$$(A.190)$$

$$\frac{Q(q^{16}, -q^6)}{(q)_\infty} = \sum_{n=0}^{\infty} \frac{q^{n(n+2)}(-q^2; q^2)_n}{(q)_{2n+2}} \qquad \textbf{(S. 123)}$$
$$(A.191)$$

A.1.36 Mod 36 Identities

A.1.36.1 Triple Products

$$\frac{f(-q^3, -q^{33})}{\psi(-q)} = \sum_{n=0}^{\infty} \frac{q^{n(n+4)}(q^6; q^6)_n(-q; q^2)_{n+1}}{(q^2; q^2)_{2n+2}(q^2; q^2)_n}$$
$$\text{(Dyson [Bai47, p. 434, Eq. (C1)]; } \textbf{S. 116}) \qquad (A.192)$$

$$\frac{f(-q^9, -q^{27})}{\psi(-q)} = \sum_{n=0}^{\infty} \frac{q^{n^2}(q^6; q^6)_n(-q; q^2)_{n+1}}{(q^2; q^2)_{2n+2}(q^2; q^2)_n}$$
$$\text{(Dyson [Bai47, p. 434, Eq. (C2)]; } \textbf{S. 115c}) \qquad (A.193)$$

$$\frac{f(-q^{12}, -q^{24})}{\psi(-q)} = \sum_{n=0}^{\infty} \frac{q^{n(n+2)}(q^3; q^6)_n(-q; q^2)_{n+1}}{(q^2; q^2)_{2n+1}(q; q^2)_n}$$
$$\textbf{(S. 110c)} \qquad (A.194)$$

$$\frac{f(-q^{15}, -q^{21})}{\psi(-q)} = 1 + \sum_{n=1}^{\infty} \frac{q^{n^2}(q^6; q^6)_{n-1}(-q; q^2)_n}{(q^2; q^2)_{2n-1}(q^2; q^2)_n}$$
$$\text{(Dyson [Bai47, p. 434, Eq. (C3)]; } \textbf{S. 114}) \qquad (A.195)$$

A.1.36.2   Quintuple Products

$$\frac{Q(q^{18}, q)}{(q^2; q^2)_\infty} = \sum_{n=0}^{\infty} \frac{q^{2n(n+1)}(q^3; q^6)_n}{(q^2; q^2)_{2n}(q; q^2)_{n+1}} \qquad ([\text{MS09}]) \tag{A.196}$$

$$\frac{Q(q^{18}, q^3)}{(q^2; q^2)_\infty} = \sum_{n=0}^{\infty} \frac{q^{2n^2}(q^3; q^6)_n}{(q^2; q^2)_{2n}(q; q^2)_n} \qquad (\text{Ramanujan [AB09, p. 104, Ent. 5.3.4]}) \tag{A.197}$$

$$\frac{Q(q^{18}, q^5)}{(q^2; q^2)_\infty} = \sum_{n=0}^{\infty} \frac{q^{2n(n+1)}(q^3; q^6)_n}{(q^2; q^2)_{2n+1}(q; q^2)_n} \qquad (\textbf{S. 124}) \tag{A.198}$$

$$\frac{Q(q^{18}, q^7)}{(q^2; q^2)_\infty} = \sum_{n=0}^{\infty} \frac{q^{2n(n+2)}(q^3; q^6)_n}{(q^2; q^2)_{2n+1}(q; q^2)_n} \qquad (\textbf{S. 125}) \tag{A.199}$$

A.2   False Theta Series Identities

Recalling that Ramanujan's theta series

$$f(a, b) := \sum_{n=-\infty}^{\infty} a^{n(n+1)/2} b^{n(n-1)/2}$$

$$= \sum_{n=0}^{\infty} a^{n(n+1)/2} b^{n(n-1)/2} + \sum_{n=1}^{\infty} a^{n(n-1)/2} b^{n(n+1)/2},$$

we define the corresponding *false theta series* as

$$\Psi(a, b) := \sum_{n=0}^{\infty} a^{n(n+1)/2} b^{n(n-1)/2} - \sum_{n=1}^{\infty} a^{n(n-1)/2} b^{n(n+1)/2}.$$

Notice that while it is always true that $f(a, b) = f(b, a)$, in most cases $\Psi(a, b) \neq \Psi(b, a)$.

L. J. Rogers [Rog17] studied q-series expansions for many instances of $f(\pm q^\alpha, \pm q^\beta)$ and $\Psi(\pm q^\alpha, \pm q^\beta)$, which he called *theta (resp. false theta) series of order* $(\alpha + \beta)/2$.

Note that a false theta series identity for the series $\Psi(\pm q^\alpha, \pm q^\beta)$ arises from the same Bailey pair as the Rogers-Ramanujan type identity with product $f(\pm q^\alpha, \pm q^\beta)/\varphi(-q)$. Thus, for reference, a designation of the form ($\textbf{F.}$ n) means that the identity is the false theta analog of ($\textbf{S.}$ n), the nth identity in Slater's list [Sla52]. Slater never wrote about false theta function identities.

A.2.0 False Theta Series Identities of Order $\frac{3}{2}$

$$\Psi(q^2, q) = \sum_{n=0}^{\infty} \frac{(-1)^n q^{n(n+1)/2}}{(-q)_n} \qquad ([\text{Rog17, p. 333, (5)}];\ \mathbf{F.\ 2}) \qquad (A.200)$$

$$= \sum_{n=0}^{\infty} \frac{q^{n(2n+1)}}{(-q)_{2n+1}} \qquad (\text{Ramanujan } [\text{AB05, p. 233, Entry 9.4.3}];\ \mathbf{F.\ 5})$$
$$(A.201)$$

$$= 2 - \sum_{n=0}^{\infty} \frac{q^{n(2n-1)}}{(-q)_{2n}} \qquad (\text{Ramanujan } [\text{AB05, p. 233, Entry 9.4.4}])$$
$$(A.202)$$

$$= \sum_{n=0}^{\infty} \frac{(-1)^n q^n}{(-q^2; q^2)_n} \qquad (\text{Ramanujan } [\text{AB05, p. 235, Entry 9.4.7}])$$
$$(A.203)$$

A.2.1 False Theta Series Identity of Order 2

$$\Psi(-q^3, -q) = \sum_{n=0}^{\infty} \frac{(-1)^n q^{n(n+1)} (-q; q^2)_n}{(q; q^2)_{n+1}(-q^2; q^2)_n} \qquad (\ [\text{MSZ09, Eq. (2.12)}];\ \mathbf{F.\ 11})$$
$$(A.204)$$

A.2.2 False Theta Series Identities of Order $\frac{5}{2}$

$$\Psi(q^4, q) = \sum_{n=0}^{\infty} \frac{(-1)^n q^{n(n+1)}}{(-q)_{2n+1}} \qquad ([\text{Rog17, p. 334 (7)}];\ \mathbf{F.\ 17}) \qquad (A.205)$$

$$\Psi(q^3, q^2) = \sum_{n=0}^{\infty} \frac{(-1)^n q^{n(n+1)}}{(-q)_{2n}} \qquad ([\text{Rog17, p. 334 (7)}]) \qquad (A.206)$$

A.2.3 False Theta Series Identities of Order 3

$$1 = \sum_{n=0}^{\infty} \frac{(-1)^n q^{n^2}}{(q; q^2)_{n+1}} \qquad ([\text{Rog17, p. 333 (4)}];\ \mathbf{F.\ 26})$$
$$(A.207)$$

$$\Psi(-q^5, -q) = \sum_{n=0}^{\infty} \frac{(-1)^n q^{n(n+1)}}{(q; q^2)_{n+1}} \qquad ([\text{Rog17, p. 333 (4)}]; \textbf{F. 22}; \textbf{F. 28})$$

$$(A.208)$$

$$\Psi(q^5, q) = \sum_{n=0}^{\infty} \frac{(-1)^n q^{3n(n+1)/2}(q)_{3n+1}}{(q^3; q^3)_{2n+1}} \qquad (\text{Dyson [Bai48, p. 9, Eq. (7.8)]})$$

$$(A.209)$$

$$\Psi(q^4, q^2) = \sum_{n=0}^{\infty} \frac{(-1)^n q^{3n(n+1)/2}(q)_{3n}(1 - q^{3n+2})}{(q^3; q^3)_{2n+1}}$$

$$(\text{Dyson [Bai48, p. 9, Eq. (7.9)]}) \qquad (A.210)$$

A.2.4 False Theta Series Identities of Order 4

$$\Psi(-q^2, -q^6) = \sum_{n=0}^{\infty} \frac{(-1)^n q^{n(n+1)/2}(-q)_n}{(q; q^2)_{n+1}}$$

$$([\text{Rog17, p. 333 (5)}]) \qquad (A.211)$$

$$\Psi(-q^6, -q^2) = \sum_{n=0}^{\infty} \frac{(-1)^n q^{n(n+1)/2}(-q^2; q^2)_n}{(q^{n+1}; q)_{n+1}}$$

$$([\text{Rog17, p. 333 (5)}]) \qquad (A.212)$$

$$\Psi(-q^5, -q^3) = \sum_{n=0}^{\infty} \frac{(-1)^n q^{n(n+1)/2}(q)_n(-q; q^2)_n}{(q; q)_{2n+1}}$$

$$(\text{Ramanujan [AB05, p. 257, Eq. (11.5.3)]}; \textbf{F. 37}) \qquad (A.213)$$

$$\Psi(-q^7, -q) = \sum_{n=0}^{\infty} \frac{(-1)^n q^{n(n+3)/2}(q)_n(-q; q^2)_n}{(q; q)_{2n+1}}$$

$$(\text{Ramanujan [AB05, p. 257, Eq. (11.5.4)]}; \textbf{F. 35}) \qquad (A.214)$$

$$\Psi(q^7, q) = \sum_{n=0}^{\infty} \frac{(-1)^n q^{2n(n+1)}(q^4; q^4)_n (q; q^2)_{2n+1}}{(q^4; q^4)_{2n+1}}$$

$$(\text{Ramanujan [AB05, p. 257, Eq. (11.5.5)]}) \qquad (A.215)$$

A.2.5 False Theta Series Identities of Order 5

$$\Psi(-q^7, -q^3) = \sum_{n=0}^{\infty} \frac{(-1)^n q^{n(n+1)/2}}{(q; q^2)_{n+1}} \qquad ([\text{Rog17, p. 333 (3)}]; \textbf{F. 45}) \quad (A.216)$$

$$\Psi(-q^9, -q) = \sum_{n=0}^{\infty} \frac{(-1)^n q^{n(n+3)/2}}{(q; q^2)_{n+1}} \quad \text{([Rog17, p. 333 (3)]; \textbf{F. 43})} \quad \text{(A.217)}$$

A.2.8 False Theta Series Identities of Order $\frac{15}{2}$

$$\Psi(q^8, q^7) - q\Psi(q^2, q^{13}) = \sum_{n=0}^{\infty} \frac{(-1)^n q^{n(3n-1)/2}(q)_n}{(q)_{2n}} \quad \text{([Rog17, p. 333 (2)])}$$

$$\text{(A.218)}$$

$$\Psi(q^7, q^8) + q\Psi(q^2, q^{13}) = \sum_{n=0}^{\infty} \frac{(-1)^n q^{n(3n+1)/2}(q)_n}{(q)_{2n+1}}$$

$$\text{([Rog17, p. 333 (2)]; \textbf{F. 62})} \quad \text{(A.219)}$$

$$\Psi(q^4, q^{11}) + q\Psi(q, q^{14}) = \sum_{n=0}^{\infty} \frac{(-1)^n q^{3n(n+1)/2}(q)_n}{(q)_{2n+1}} \quad \text{([Rog17, p. 333 (2)])}$$

$$\text{(A.220)}$$

A.2.9 False Theta Series Identities of Order 9

$$\Psi(q^{15}, -q^3) = \sum_{n=0}^{\infty} \frac{(-1)^n q^{n(n+3)/2}(-q^3; q^3)_n}{(q; q^2)_{n+1}(-q)_n(-q)_{n+1}} \quad \text{([MS08, p. 768, Eq. (1.32)])}$$

$$\text{(A.221)}$$

$$\Psi(q^{12}, -q^6) + q^2\Psi(q^{18}, -1) = \sum_{n=0}^{\infty} \frac{(-1)^n q^{n(n+1)/2}(-q^3; q^3)_n}{(q; q^2)_{n+1}(-q)_n^2}$$

$$\text{([MS08, p. 768, Eq. (1.34)])} \quad \text{(A.222)}$$

$$\Psi(q^{15}, q^3) = \sum_{n=0}^{\infty} \frac{(-1)^n q^{n(n+3)/2}(q^3; q^3)_n}{(1 + q^{n+1})(q)_{2n+1}}$$

$$\text{(Dyson [Bai47, J6; p. 434, Eq. (E1)]; \textbf{F. 76})} \quad \text{(A.223)}$$

$$= \sum_{n=0}^{\infty} \frac{(-1)^n q^{n(n+3)/2}(q^3; q^3)_n(1 - q^{n+1})}{(q)_{2n+2}}$$

$$\text{([MSZ09, Eq. (2.13)]; \textbf{F. 75})} \quad \text{(A.224)}$$

$$\Psi(q^{12}, q^6) = \sum_{n=0}^{\infty} \frac{(-1)^n q^{n(n+1)/2}(q^3; q^3)_n}{(q)_{2n+1}}$$

$$\text{(Dyson [Bai47, p. 434, Eq. (E2)]; \textbf{F. 77})} \quad \text{(A.225)}$$

A.2.11   False Theta Series Identities of Order $\frac{21}{2}$

$$\Psi(q^8, q^{13}) + q^2\Psi(q, q^{20}) = \sum_{n=0}^{\infty} \frac{(-1)^n q^{n(n+1)/2}(q)_n}{(q)_{2n+1}}$$

([Rog17, p. 332 (1)]; F. 80)   (A.226)

$$\Psi(q^{10}, q^{11}) - q\Psi(q^4, q^{17}) = \sum_{n=0}^{\infty} \frac{(-1)^n q^{n(n+1)/2}(q)_n}{(q)_{2n}}$$

([Rog17, p. 332 (1)]; F. 81)   (A.227)

$$\Psi(q^5, q^{16}) + q\Psi(q^2, q^{19}) = \sum_{n=0}^{\infty} \frac{(-1)^n q^{n(n+3)/2}(q)_n}{(q)_{2n+1}}$$

([Rog17, p. 332 (1)]; F. 82)   (A.228)

A.2.16   False Theta Series Identities of Order 16

$$\Psi(-q^8, -q^{24}) = \sum_{n=0}^{\infty} \frac{(-1)^n q^{n(n+3)/2}(q)_{n+1}(-q^2; q^2)_n}{(q)_{2n+2}}$$

([MSZ09, Eq. (2.14)]; F. 103)   (A.229)

$$\Psi(q^{22}, q^{10}) + q\Psi(q^{26}, q^6) = \sum_{n=0}^{\infty} \frac{(-1)^n q^{n(n+3)/2}(q)_n(-q; q^2)_n}{(q)_{2n+1}}$$

([MSZ09, Eq. (2.15)]; F. 106)   (A.230)

A.2.18   False Theta Series Identities of Order 18

$$\Psi(q^{21}, -q^{15}) - q\Psi(q^{27}, -q^9) = \sum_{n=0}^{\infty} \frac{(-1)^n q^{n(n+1)}(-q^3; q^6)_n}{(q^2; q^4)_n(-q; q)_{2n+1}}$$

([MS08, p. 768, Eq. (1.31)])   (A.231)

$$\Psi(q^{21}, -q^{15}) + q^3\Psi(q^{33}, -q^3) = \sum_{n=0}^{\infty} \frac{(-1)^n q^{n(n+1)}(-q^3; q^6)_n}{(q^2; q^4)_{n+1}(-q; q)_{2n}}$$

([MS08, p. 768, Eq. (1.33)])   (A.232)

$$\Psi(q^{27}, -q^9) + q^2\Psi(q^{33}, -q^3) = \sum_{n=0}^{\infty} \frac{(-1)^n q^{n(n+3)}(-q^3; q^6)_n}{(q^2; q^4)_{n+1}(-q; q)_{2n}}$$

([MS08, p. 768, Eq. (1.35)])   (A.233)

$$\Psi(q^{21},q^{15}) + q\Psi(q^{27},q^9) = \sum_{n=0}^{\infty} \frac{(-1)^n q^{n(n+1)}(q^3;q^6)_n}{(q^2;q^4)_n(-q^2;q^2)_n(q;q^2)_{n+1}}$$

([MS08, p. 768, Eq. (1.36)]) (A.234)

$$\Psi(q^{21},q^{15}) - q^3\Psi(q^{33},q^3) = \sum_{n=0}^{\infty} \frac{(-1)^n q^{n(n+1)}(q^3;q^6)_n}{(q^{2n+2};q^2)_{n+1}(q;q^2)_n}$$

([MS08, p. 769, Eq. (1.38)]; **F. 107**) (A.235)

$$\Psi(q^{27},q^9) - q^5\Psi(q^{39},q^{-3}) = \sum_{n=0}^{\infty} \frac{(-1)^n q^{n(n+3)}(q^3;q^6)_n}{(q^{2n+2};q^2)_{n+1}(q;q^2)_n}$$

([MS08, p. 769, Eq. (1.40)]) (A.236)

Appendix B

Every attempt has been made to transcribe the letters contained herein as accurately as possible from the originals. Accordingly, grammatical errors and crossed out words in the original are retained here.

B.1   Letters from W. N. Bailey to F. J. Dyson

Freeman Dyson corresponded with W. N. Bailey in late 1943 and 1944. Dyson was twenty years old at the time. Dyson preserved the letters he received from Bailey and was kind enough to allow them to be transcribed for use in this volume. In these letters, we obtain insight into Bailey's thoughts and struggles as he worked on the two papers [Bai47, Bai48] that solidified the ideas of what came to be known as Bailey pairs and Bailey's lemma, as well as set the stage for much future research into Rogers–Ramanujan type identities.

B.1.1   Bailey to Dyson, December 22, 1943

17 Prince's Road
Heaton Moor,
Stockport,
Cheshire

Dec. 22 / 43.

Dear Dyson,

I am writing to you in connection with my paper for the L.M.S. I had a letter from Hardy two days ago, & he enclosed your report (a rather unusual thing to do, but very useful in the circumstances).

Many thanks for your comments. Actually I have not seen Rogers' papers for some years & I am in the unfortunate position of not being able to look them up. The only copy in Manchester (as far as I am aware) was destroyed in the blitz in 1940, or I should certainly have refreshed my memory. I am now trying to purchase the parts of the *L.M.S. Proceedings* in which they appear, but don't know whether I shall be successful. I shall, of course, alter the last part of §1 and the first part of §2 to meet your criticism.

With regard to the formulæ (4.3) & (5.3), I did not remember that they were given by Rogers and I couldn't look up this paper either. I had, however, discovered them in a paper by Jackson, a copy of which I enclose as it might interest you. [He sent me two copies.] Jackson makes no reference to Rogers. The formulæ are the last on p. 175 & the first on p. 176. (with a misprint). My (6.3) was, I thought, new but I think the third formula on p. 170 of Jackson's paper is meant to be the same, but it is wrong. I am not sure that he deserves to be quoted when he simply states the formula & gives it incorrectly, but I shall probably put in a reference. Jackson is terribly careless & has caused me a good deal of trouble, but he has a good many curious results in this paper.

With regard to your formulæ for products in which the powers of x advance by 27, I am afraid I don't see how you obtained them. I should very much like to know. The products in the first three are the products occurring in my formulæ for 9's with x replaced by x^4, as I suppose you noticed. If you like you could make a short paper about them & ask for it to follow mine probably in the *Proceedings* or write one for the *Journal* quoting what is necessary of my paper. If you still don't think it worth while making a separate paper, I will incorporate the formulæ in my paper with due acknowledgements. I should however have to give at least some indication of how they were obtained. I may be dense, but they don't seem at all obvious to me. I shall certainly be interested to know how you got them.

Yours sincerely,

W. N. Bailey

B.1.2   Bailey to Dyson, December 24, 1943

17 Prince's Road
Heaton Moor,
Stockport,
Cheshire
Dec. 24/43

Dear Dyson,

Just a line to tell you not to bother writing out the proofs of your identities—if I am not already too late. I have rather belatedly found out how you got them.

Two of them come from

$$1 + \sum_{n=1}^{\infty} \frac{(-1)^n [ax^3]_{n-1}(1 - ax^{6n}) a^{4n} x^{\frac{1}{2}(27n^2 - 3n)}}{[x^3]_n}$$

$$= \prod_{1}^{\infty}(1 - ax^m) \sum_{n=0}^{\infty} \frac{a^n x^{n^2} [ax^3]_{n-1}}{x_n!(ax)_{2n-1}}$$

where [ ] denote that powers of x advance by 3. Two of your identities come by taking $a = 1$ & $a = x^3$. I suppose you got your results in the same way as I have done, by taking $\alpha_r = 0$ unless r is a multiple of 3. I should also have got the last ~~third~~ formula of the original three, but something has gone wrong & I cannot found the error at present. Perhaps it is too near Xmas!

I think several other formulæ should be obtainable by similar methods, but of course they may not be new.

By the way, the third formula on p. 170 of Jackson's paper is evidently meant to be one of mine. All that is needed to put it right is to change the sign of q on one side.

If you decide to let your identities go in my paper I will send the paper on to you when finished. They would put a finishing touch on my paper, & definitely increase its value, but I think you would be perfectly justified in making a separate paper. Best seasonal greetings.

W. N. Bailey

B.1.3 Bailey to Dyson, January 5, 1944

17 Prince's Road
Heaton Moor,
Stockport,
Cheshire
Jan. 5/44

Dear Dyson,

I am writing to give you information this time—not to worry you.

Your method of getting the "27" identities appears to be equivalent to the method I found. I have added two paragraphs to my paper, the first bringing in the idea of making various $\alpha_r = 0$ (or taking various $\cos\theta = 0$, & getting in particular two of your "27" identities. In the last paragraph, I give a list of all the identities you sent, but I ~~didn't~~ don't attempt to provide proofs. Actually I have only looked at Rogers' methods from one aspect, & I doubt whether proofs of all your identities could be got by the methods of my paper.

I have now got a copy of Rogers' paper which he wrote in 1917—borrowed it from Jackson who very kindly sent me some others including Rogers' "Third Memoir", but unfortunately not the other two.

By the way, the formula for $\prod_1^\infty \left(\frac{1-x^{9n}}{1-x} \right)$ [sic] is a brute to get. It is easy enough to get a series for it, but I find it an awful business to get your series.

I thought it wasn't necessary to worry you again with my paper, particularly as you are so busy, but I wanted to let you know that I have put your identities on record.

Yours sincerely,
W. N. Bailey

B.1.4   Bailey to Dyson, February 13, 1944

17 Prince's Road
Heaton Moor,
Stockport,
Ches.
Feb. 13/44

Dear Dyson,

I was interested in your last letter which I received some time ago. I should think your proof of the identity for $\prod_1^\infty (1 - x^{9n})$ is as short as can be expected. I have got an a-generalisation for it, but unfortunately it is not at all elegant. It is, in fact,

$$1 + \sum_1^\infty (-1)^n \frac{\{ax^3\}_{n-1}}{x_n^3!} (1 - ax^{6n})(1 - x^{3n} + ax^{6n}) a^{4n-1} x^{\frac{1}{2}(27n^2 - 9n)}$$

$$= \prod_{m=1}^\infty (1 - ax^m) \sum_{n=0}^\infty \frac{\{ax^3\}_n}{x_n!(ax)_{n+1}} a^n x^{n^2 + n},$$

where $\{a\}_n = (1 - a)(1 - ax^3) \cdots (1 - ax^{3n-3})$, and I don't see any nicer way of writing it.

I am really writing to you to let you know how things are going. I have been pretty busy in other ways lately, but I have done enough to feel rather disappointed with this sort of thing. After studying Rogers' papers, I was led to put things in this way: If $\beta_n = \sum_{r=0}^n \alpha_r u_{n-r} v_{n+r}$, and $\gamma_n = \sum_{r=n}^\infty \alpha_r u_{r-n} v_{r+n}$, then $\sum_{n=0}^\infty \alpha_n \gamma_n = \sum_{n=0}^\infty \beta_n \delta_n$, provided of course that convergence conditions are satisfied. This leads, in particular, to all the known transformations of ordinary hypergeometric series. In fact, it is substantially equivalent to the method used in my tract, though I think this is rather a more illuminating way of putting it. Similarly it is substantially equivalent to the method I used in my last paper to find the transformation of a nearly poised basic series. One form we can take is

$$\beta_n = \sum_{r=0}^n \frac{\alpha_r}{(q)_{n-r}(aq)_{n+r}}$$

$$\gamma_n = \sum_{r=n}^\infty \frac{\delta_r}{(q)_{r-n}(aq)_{r+n}}$$

& then

$$(aq)_{2n}\beta_n = \text{Rogers' } w_{2n} \text{ if } a = 1$$

$$= \text{ " } w_{2n+1} \text{ if } a = q, \text{ while}$$
$$\delta_n = \alpha_{2n} \text{ if } a = 1$$
$$= \alpha_{2n+1} \text{ if } a = q$$

and $\gamma_n = b_{2n}$ or b_{2n+1}, $\alpha = v_{2n}$ or v_{2n+1}.

The formula for γ_n ~~is equivalent to~~ gives, if $\delta_r = (\rho_1)_r(\rho_2)_r(aq/\rho_1\rho_2)^r$, by Gauss's theorem,

$$\gamma_n = \frac{(\rho_1)_n(\rho_2)_n}{(aq/\rho_1)_n(aq/\rho_2)_n} \left(\frac{aq}{\rho_1\rho_2}\right)^n \cdot \prod_{m=1}^{\infty} \frac{(1 - aq^m/\rho_1)(1 - aq^m/\rho_2)}{(1 - aq^m)(1 - aq^m/\rho_1\rho_2)}$$

Consequently

$$\sum_{n=0}^{\infty} (\rho_1)_n(\rho_2)_n(aq/\rho_1\rho_2)^n \beta_n$$

$$= \prod_{m=1}^{\infty} [\ ] \sum_{n=0}^{\infty} \frac{(\rho_1)_n(\rho_2)_n}{(aq/\rho_1)_n(aq/\rho_2)_n} \left(\frac{aq}{\rho_1\rho_2}\right)^n \alpha_n.$$

When $a = 1$ or $a = q$ this is equivalent to Rogers's formulæ with u & v in them, from which he deduces 19 particular cases giving Fourier series in terms of A's. Actually $u = q^{\frac{1}{2}}/\rho_1$, $v = q^{\frac{1}{2}}/\rho_2$.

Similarly the relations between β_n & α_n gives, with the analogue of Dougall, relations corresponding to those given by Dougall connecting the a's and b's. It is evident from all this, for example, that any results obtained from Rogers' formulæ E1, E3, F1, F2, E2, E7, F3, F4, (& perhaps other) all the 19 uv formulæ can be derived from Watson's transformation directly.

Of course, one could work out a few formulæ which would generalise all those given by Rogers or obtainable by the methods of his 1917 paper, but, apart from the results given already, these formulæ appear to be anything but elegant, some have some factors advancing by $\sqrt{q}$ & some by q, & the series are not of any general type. In fact it seems to me that, apart from the general transformations already given, the method is only useful for obtaining formulæ of the Rogers–Ramanujan type. They are, at any rate, reasonably simple in appearance.

Of course the simple result at the beginning of this letter has its analogue for integrals. Thus if

$$F(y) = \int_0^y \phi(x)f(y - x)g(y + x)\, dx,$$

and

$$G(y) = \int_y^{\infty} \psi(x)f(x - y)g(x + y)\, dx,$$

then

$$\int_0^{\infty} \phi(x)G(x)\, dx = \int_0^{\infty} \psi(x)F(x)\, dx.$$

One might hope that results could be got for integrals corresponding to those got for series, but the trouble is to start. So far I have got nowhere.

The result of all this is that I feel that the only thing I am being led to is a search for more R-R identities, & probably you have found the most interesting ones that are new. Of course the method gives a-generalisations of them.

Yours sincerely,

W. N. Bailey

B.1.5   Bailey to Dyson, August 1, 1944

17 Prince's Road
Heaton Moor,
Stockport,
Cheshire,
Aug 1/44.

Dear Dyson,

Hardy has passed on your comments on my paper on "Identities of the Rogers–Ramanujan type." Many thanks for reading it so carefully & for finding the errors. I had checked the formulæ a little, but evidently not enough. The first two formulæ were got from

$$1 + \sum_{n=1}^{\infty} \frac{(\rho_1)_n (\rho_2)_n \{ax^3\}_{n-1}}{x_n!(ax)_{2n-1}} \left(\frac{ax}{\rho_1 \rho_2} \right)^n$$

$$= \prod_{m=1}^{\infty} \left[\frac{(1-ax^m/\rho_1)(1-ax^m/\rho_2)}{(1-ax^m)(1-ax^m/\rho_1\rho_2)} \right]$$

$$\times \left[1 + \sum_{n=1}^{\infty} \frac{(-1)^n \{ax^3\}_{n-1}(1-ax^{6n})(\rho_1)_{3n}(\rho_2)_{3n}}{x_n^3!(ax/\rho_1)_{3n}(ax/\rho_2)_{3n}} \right.$$

$$\left. \times \frac{a^{4n} x^{\frac{3}{2}n(3n+1)}}{\rho_1^{3n} \rho_2^{3n}} \right] \quad (6.4)$$

where $\{a\}_n = (1-a)(1-ax^3)\cdots(1-ax^{3n-3})$. If we take $\rho_1 = -\sqrt{a}$, $\rho_2 = -\sqrt{ax}$, this becomes

$$1 + \sum_{n=1}^{\infty} \frac{(-\sqrt{a})_n (-\sqrt{ax})_n \{ax^3\}_{n-1} x^{\frac{1}{2}n}}{x_n!(ax)_{2n-1}}$$

$$= \prod_{m=1}^{\infty} \left[\frac{(1+\sqrt{a}x^m)(1+\sqrt{a}x^{m+\frac{1}{2}}))}{(1-ax^m)(1-x^{m-\frac{1}{2}})} \right]$$

$$\times \left[1 + \sum_{n=1}^{\infty} \frac{(-1)^n \{ax^3\}_{n-1}(1-\sqrt{a}x^{3n})(1+\sqrt{a})a^n x^{9n^2}}{x_n^3!} \right]$$

I got the first incorrect result by taking $\sqrt{a} = x^{\frac{3}{2}}$. I find that I dropped a factor of $(-1)^n$, & the formula I now get is

$$\sum_{n=0}^{\infty} \frac{x_n^6! x^n}{x_{2n+1}! x_n^2!} = \prod_{n=1}^{\infty} \frac{(1 - x^{18n})(1 - x^{18n-3})(1 - x^{18n-15})}{(1 - x^n)(1 - x^{2n-1})} \tag{7.1}$$

Similarly by taking $\sqrt{a} = 1$, I got the second formula, viz.

$$1 + 2 \sum_{n=1}^{\infty} \frac{x_{n-1}^6! x^n}{x_{2n-1}! x_n^2} = \prod_{n=1}^{\infty} \frac{(1 + x^n)(1 - x^{9n})}{(1 - x^n)(1 + x^{9n})}. \tag{7.2}$$

You say this is ~~wrong~~ correct up to the term x^6. I must confess that I cannot find anything wrong in the working.

The third formula was certainly wrong, & should have been the same as the first of the five you sent. I will incorporate these formulæ ~~you~~ in the paper, but I should be very glad if you could say whether you agree with (7.1) & (7.2) now. I find these things very tedious to check to any extent, though I thought I was fairly safe.

Again, many thanks for all the care you have taken & for the new results.

Yours sincerely,

W. N. Bailey

B.1.6 Bailey to Dyson, October 8, 1946

8 Langton Avenue,
Wetstone,
London, N. 20.
Oct. 8/46

Dear Dyson,

I was interested to hear from you again & that you are back in Cambridge. I am now in London (at Bedford College) & have been for the past two years.

I have had two papers ready for the P.L.M.S. for about 3 years or more, so they ought to be published in another year or two! Actually the L.M.S. have done all they can to speed up publication, but first of all shortage of paper & then shortage of labour have been too much for them.

After coming here I had a rather strenuous time getting used to the ways of London University, finding a house, & so on, so I didn't make much progress with the work I was doing.

Lately, however, I have sent a paper to the Quarterly Journal which I give the basic analogues of 6.6(3) in my tract, & of 6.8(3) & 7.6(2). The first of these is what Dougall's theorem becomes when the series does not terminate. The other two are the relations connecting 4 $_9F_8$'s. I found the analogue of 6.6(3) was merely a particular case of a formula given in a Q.J. paper in 1936 (Series of hyp. type infinite in both dirs., Q.J. 7 (105) first formula in §5).

I got the analogue of 6.8(3) by transforming the arguement in §§6.7& 6.8 of my tract into series by considering poles on the right of the contours. Then I did the corresponding work for basic series. The idea was simple enough, but the details nearly broke my heart.

In P.L.M.S. 42 (1937) 410–421, Whipple gave (or rather showed how to find) a connection between 4 $_9F_8$'s (well-poised) when there was no restriction on the sums of numerator & den. parameters. This leaves the obvious problem of finding the corresponding result for basic series, but I hadn't the pluck to start that. Whipple's proof is very short & depends on a contour integral, but I don't see how one could adapt his method to basic series, unless one worked out a good deal about integrals generalising integrals of Barnes' type.

There was another thought I had that seemed to hold promise at one time, but I never got anything out of it. In the papers you saw I gave a general theorem on series which has the integral analogues:

$$\text{If } F(y) = \int_0^y \phi(x)f(y-x)g(y+x)\ dx$$

$$\&\ G(y) = \int_y^\infty \psi(x)f(x-y)g(x+y)\ dx,$$

then

$$\int_0^\infty \phi(x)G(x)\ dx = \int_y^\infty \psi(x)F(x)\ dx.$$

With so much being derivable from the series theorem, I thought that there might be possibilities from the integral theorem, but I did not succeed in getting anything interesting. Still, there may be something in it.

I didn't try to get any more identities of the Rogers–Ramanujan type. We got a good many between us 3 years ago!

Yours sincerely,

W. N. Bailey.

B.2   Letter from F. H. Jackson to Dyson, June 4, 1944

35(?) Osbourne Road
Eastbourne

Dear Sir,

My friend Dr W N Bailey tells me you are interested in q-analysis so I venture to send you some of my old papers. They may interest you and probably suggest new lines of work which I am too old either to see or undertake At 74

I find any long continued effort beyond me. At the moment I am trying with little success to get basic hypergeometric series into a workable

$$F(a, b; c, x)\{1 + f_1(a, b, c)\epsilon + f_2(a, b, c)\epsilon^2 + \cdots \text{ ad inf} = \Phi(a, b; c, q, x)$$

ϵ small and $\sum f_n(a, b, c, x)\epsilon^n$ convergent. It looks easy but the f's are awkward even for ϵ^3. The late Prof. Forsyth suggested this problem to me 40 years ago.

I am out of spare copies of papers on q-integrations: reversing

$$\frac{q^\theta - 1}{x(q - 1)}$$

but as Forsyth once said to me, too much reading of other folks works has its dangers to originality: to set against its saving one from rediscovery of old theorems: As he crudely put it—more folk are choked to death than starved to death. Choking is our danger in Cambridge.

With best wishes
Yours sincerely
(Rev & Canon) F. H. Jackson

P. S. Please forgive my ignorance of your Christian name implied by the address "Dyson, Esq".

B.3   Letters from L. J. Rogers to F. H. Jackson

A pair of letters from L. J. Rogers to F. H. Jackson dated February 13th and 14th (with no year specified, but probably 1917), remain extant. These letters were given to Lucy Slater by F. H. Jackson's sister upon Jackson's death in 1960. Slater preserved them for a number of years, before passing them along to Harold Exton, who eventually passed them on to George Andrews at some point during the 1990s. Andrews published the entire contents of the letters, along with commentary, in [And82], and this paper has been reproduced in [And13b, pp. 977–991]. He published the contents of these two letters once again in [And13a]. Accordingly, we do not reproduce the letters here. These letters contain some interesting mathematics that foreshadow later work in q-series, and also reveal Rogers' apparent casual attitude about having had his work ignored by the mathematical community for nearly a quarter-century only to have (2.3) and (2.4) rise to prominence as a conjecture attributed to Ramanujan!

B.4   Letters from W. N. Bailey to L. J. Slater

In the following correspondence, $(q; n) := (q; q)_n$ and $(a; q, n) := (a; q)_n$.

B.4.1   Bailey to Slater, August 2, 1950

8 Langton Avenue
Whetstone, N. 20
Aug 2/50

Dear Miss Slater,

Many thanks for the list of results which I received yesterday. I am glad to have them.

The third result (with 20) on page xiii can be simplified, & it would be a pity not to do so. The series on the right is $\sum \frac{q^{n^2}}{(q;2n)}$.

I enclose a little paper which I sent to the L. M. S. about a fortnight ago, on your work. I hope you will be able to make it out, but it is all I have got. (1.4) & (1.5) are two of the four results given by Watson in his Q. J. paper, but all 4 were actually given by Rogers. (1.6) & (1.7) are taken from your paper already accepted. (1.8) is from the prvious list you gave me, & so is (4.2). You don't seem to have given (4.2) in your new list, but you give (4.3) on p. (iii) with the sign of q changed (from G.4). I see also that you give (3.2) in your new list, the last formula on p. vi.

Of course I have only given a few examples as illustrations, so I don't think I have queered your pitch at all. The statement at the end of §2 was true of your old list. I hope it is still true! In any case there is room for a new chapter in your thesis, in which you simplify (if possible) all the formulæ which contain two products. ~~e.g.~~ [For the products in the second formula on p. xv, we get from (2.1)

$$\prod(1 - q^{27n-15})(1 - q^{27n-12})(1 - q^{27n}) + q \prod(1 - q^{27n-21})()() $$

& I don't think that (2.1) is any help here. This doesn't matter as we have got series for 27 with single products.]

I should have thought that (2.1) would be a particular case of the formulæ connecting products of four ϑ-series, but I have not been able to get it from any of those.

Don't bother to return my paper until next term. I thought it was worth writing up, as I had always that the 21 formulae could not be simplified & it seems that they can.

Many thanks for the news about Miss Jackson. She has enough to keep her busy for a considerable time!

Hope you have a good time in Oldham. I was away from home last week, & we go to Norway in a little over a fortnight.
Yours sincerely,
W. N. Bailey

Commentary. At the time of this letter, Lucy Slater was finishing up her PhD under Bailey's supervision at Bedford College, University of London. Slater was the first of a total of three PhD students supervised by Bailey. Slater was awarded her degree in 1951, with a thesis entitled *Functions of Hypergeometric Type* [Sla51a]. "Miss [Margaret] Jackson," referred to in the penultimate paragraph was Bailey's second student, who completed her PhD at Knottingham in 1952. His final student was Ratan Agarwal (PhD, London, 1953).

Slater published a condensed version of her thesis as [Sla51b, Sla52]. The appendix of her thesis (27 pages numbered in Roman numerals) and her second published paper [Sla52] contain the famous list of 130 identities of Rogers–Ramanujan type. The "little paper which ... [Bailey] sent to the L. M. S. about a fortnight ago" is [Bai51]. The interested reader will need to cross reference this letter to [Bai51] in order to see what equations the numbers refer to. The main point of [Bai51] is to restate, and give a simple proof of, the quintuple product identity (which Bailey attributes to Watson as he was unaware of the earlier discoveries by Fricke and Ramanujan) in a different form. This alternate form makes it easier to simplify certain expressions that appear as weighted sums of infinite products in some of Slater's identities into infinite products. Additionally, he explicitly states a "2-dissection" of a theta function [Bai51, p. 220, Eq. (4.1)], which we may state succinctly here in Ramanujan's theta-function notation as

$$f(z^2 q, z^{-2} q^3) + z f(z^2 q^3, z^{-2} q) = f(z, q/z). \tag{B.1}$$

The point is that many of the identities deduced by Slater via inserting specific Bailey pairs into limiting cases of Bailey's lemma, naturally initially led to a series equal to the quotient of a weighted sum of two series divided by a theta series. Bailey points out that many of these weighted sums of two theta series can be simplified to a single theta series via either the quintuple product identity or (B.1).

For example, Bailey points out that Slater found the identity

$$\sum_{n=0}^{\infty} \frac{q^{3n^2}}{(q^4; q^4)_n (q^2; q^2)_n} = \frac{f(-q^9, -q^{11}) - q^2 f(-q, -q^{19})}{(q^2; q^2)_\infty}, \tag{B.2}$$

where $f(a, b)$ is once again Ramanujan's theta function notation. Bailey shows that the right-hand side may be simplified so that the identity becomes

$$\sum_{n=0}^{\infty} \frac{q^{3n^2}}{(q^4; q^4)_n (q^2; q^2)_n} = \frac{(-q; q^2)_\infty}{(-q; -q^5)_\infty (q^4; -q^5)_\infty}. \tag{B.3}$$

What Bailey overlooked at the time was that Eq. (B.3) actually appears in Rogers' 1894 paper, albeit with q replaced by $-q$ [Rog94, p. 339, Ex. 2]. Apparently this was noticed later, as neither (B.2) nor (B.3) made it into Slater's thesis, and Rogers' version appears as Eq. (19) in Slater's list.

In any event, Slater did not include the product simplifications suggested by Bailey here in her thesis, although they were incorporated in the published version [Sla52].

B.4.2 Bailey to Slater, September 5, 1950

8 Langton Avenue
Whetstone, N. 20
Sept 5/50

Dear Miss Slater,

Thank you very much for your good wishes & for the beautiful tie you sent me. My wife says that it is the best tie I have ever had! It came as a great surprise as I had no idea that you knew my birthday was on the 5th.

I was very interested in your deduction of my (2.1) from the previous result. I have gone through the details—easy enough when you are told what to do—& I wonder how you got it. I must add a note to my paper, or perhaps it would be better still if you wrote a short note to follow my paper, but we had better wait to see if my paper is accepted before doing anything about it.

I agree that your collection of results should be sent to the L. M. S. with a short introduction. I think they certainly ought to be published. I should like to point out a few things while I remember. On p. (ii), at the foot of the page $1 + \sum_{n=1}$ can be written $\sum$. The same is true on p. (iv) (B1) & you want a bracket at the end. On p. (xiii) $\sum \frac{(-q;q^2,n)q^{n^2}}{(q^2;q^4,n)(q^2;q^2,n)}$ (C1) can be written more simply as $\sum \frac{q^{n^2}}{(q;2n)}$. On p. (xiv), 3rd set (A7), $1 + \sum_1^\infty$ can be written $\sum$. Similarly on p. (xvi), first set (A7), $1 + \sum_1^\infty$ can be written $\sum$. On p. xx I have actually found a mistake! In the last formula q^{n^2+4n-1} should be q^{n^2+4n}. Did I mention some of these before?

We had a fine time in Norway. The weather was a bit mixed, but it wasn't too bad on the whole.

Again, very many thanks for your very nice present.
Yours sincerely,
W. N. Bailey

Commentary. At the time of this letter, Slater is putting the finishing touches on her London PhD thesis. Although the copy of the thesis extant in George Andrews's collection of Slater's papers is undated, according to [Abb01], Slater received her London PhD in 1951, and went on to complete her Cambridge PhD with a thesis entitled *Hypergeometric Integrals* in 1953.

In fact, Slater earned seven degrees in all: BA General, BA Honors, MA, PhD, and DLitt at London, as well as a PhD and an ScD at Cambridge (!) [Abb01]

B.4.3 Bailey to Slater, September 6, 1953

8, Langton Avenue,
Whetstone, N. 20.
Sept 6/53.

Dear Miss Slater,

Thank you very much for the parcel which I received yesterday. It was awfully good of you to send the hankies which are very nice indeed, & they will be very useful. I was also very pleased to have the copy of the thesis & the reprints of your paper.

Yesterday we celebrated the occasion by going to Whipsnade Zoo. We had never been there before & we enjoyed it very much, especially as it was a glorious day.

Probably you remember the formula of Ramanujan's

$$x\{(1-x)(1-x^2)\cdots\}^3\{(1-x^7)(1-x^{14})\cdots\}^3$$
$$+\frac{8x^2\{(1-x^7)(1-x^{14})\cdots\}^7}{(1-x)(1-x^2)\cdots}=$$
$$=\frac{x(1+x)}{(1-x)^3}+\frac{x^2(1+x^2)}{(1-x^2)^3}-\frac{x^3(1+x^3)}{(1-x^3)^3}+\frac{x^4(1+x^4)}{(1-x^4)^4}$$
$$-\cdots-\cdots+\cdots.$$

I spent a good deal of time trying to prove this some time ago, but without success. Now N. J. Fine of Princeton has proved it. If

$$f(x)=x^2\prod_{m\geq1}\frac{(1-x^{7m})^7}{(1-x^m)}$$
$$f^*(x)=x\prod_{m\geq1}(1-x^m)^7\sum_{n\geq0}p(7n+5)x^n,$$
$$S(x)=\sum_{N\geq1}x^NN^2\sum_{d|N}\frac{\chi(d)}{d^2},\qquad\chi(d)=\left(\frac{d}{7}\right),$$
$$Q(x)=x\prod_{m\geq1}(1-x^{7m})^3(1-x^m)^3,$$

Ramanujan showed that

$$f^*(x)=7Q(x)+49f(x).\tag{1}$$

Fine has now proved that

$$f^*(x)=7\{S(x)-f(x)\}.\tag{2}$$

From (1) & (2) we get

$$S(x) = Q(x) + 8f(x) \tag{3}$$

which is the identity we wanted to prove. This shows that (3) is true. Of course what we really wanted was a proof of (3) from which (1) & (2) could be easily got. Anyway, I find Fine's work interesting.

Thanks again very much for the parcel. May I also wish your mother many happy returns.

Yours very sincerely,
W. N. Bailey

B.4.4   Bailey to Slater, December 11, 1953

8, Langton Avenue,
Whetstone, N.20.
Dec. 11/53.

Dear Miss Slater,

Many thanks for the reprints & for the very amusing Christmas card. I haven't had your paper about the $_1\Phi_1$ from the C. P. S., but I got another one from them about products & a table of $1/\prod(1-aq^n)$. You will be getting this back with some comments, so you will recognise my scrawl. Your paper reminded me of §4 of the enclosed paper, but I don't think the things are the same, though there is a fairly strong family resemblance in the method.

I suggested that you should add a footnote to say that N. J. Fine (Institute for Advanced Study, Princeton, N. J.) has proved one of the "7" formulæ , but it has not yet been published. He writes

$$f(x) = x^2 \prod_{m\geq 1} \frac{(1-x^{7m})^7}{1-x^m} = \prod_{m\geq 1}(1-x^{7m})^7 \sum_{n\geq 2} p(n-2)x^n,$$

$$f^*(x) = x \prod_{m\geq 1}(1-x^m)^7 \sum_{n\geq 0} p(7n+5)x^n,$$

$$S(x) = \sum_{N\geq 1} x^N N^2 \sum_{d|N} \frac{\chi(d)}{d^2}, \qquad \chi(d) = \left(\frac{d}{7}\right),$$

$$Q(x) = x \sum_{m\geq 1}(1-x^{7m})^3(1-x^m)^3.$$

Then Ramanujan's published identity is equivalent to

$$f^*(x) = 7Q(x) + 49f(x). \tag{1}$$

N. J. Fine has proved that

$$f^*(x) = 7\left\{S(x) - f(x)\right\}. \tag{2}$$

Combining (1) & (2),
$$S(x) = Q(x) + 8f(x), \tag{3}$$
& this is exactly one of the "7" formulæ except for a "trivial transformation" of $S(x)$.

Fine was hoping to get the other identity, but he had not tried it when he wrote to me in July.

Of course we should have liked a direct proof of (3) from which (1) & (2) can easily be derived, but Fine has at least proved that (3) is true. Of course his (2) gives a proof that $p(7n + 5)$ is divisible by 7, different from Ramanujan's.

I was interested in your news of F. H. Jackson. Also in your remarks about Searle. He was a character at Cambridge for a long time & I have heard a lot about him, but never met him. I had thought that he must have passed on by now.

Glad you are enjoying being a Fellow. Perhaps your next college will be "St. Trinians".

I have had some correspondence lately with G. N. Watson. He has sent a paper to the new Glasgow journal about

$$I_n \equiv \int_0^1 x^\alpha (1-x)^\beta \frac{d^n}{dx^n} \left\{ x^\gamma (1-x)^\delta \right\} dx,$$

& sent me a carbon copy. The formula he got was a relation between three series $_3F_2(1)$, so I am going (in time) to send a note to Glasgow about contiguous functions of the type $_3F_2(1)$. Watson has sent me another manuscript & says gleefully "I defy you to get anything about $_3F_2$'s out of the attached typescript". Again, many thanks & best wishes to your mother & yourself.

Yours sincerely,
W. N. Bailey.

B.4.5 Bailey to Slater, September 6, 1961

1 Baldwin Avenue,
Eastbourne.
Sept 6/61

Dear Miss Slater,

Many thanks for your interesting letter & for your very kind gift of the socks, which will be very useful. You sent me a pair of a similar kind a few years ago, & since then I have always used the same kind.

I was glad to hear you were getting on so well with the new book, and that "Conf. hyp. fns" is bringing its financial reward.

An American firm is reprinting the earlier Cambridge Tracts, including my own, by a photographic process. They pay royalties, with quite a nice payment in advance.

I am sorry to say that I have had a very poor summer. In May I was suddenly taken ill. I was sent to hospital "for a week or ten days" but eventually I was there for 5 weeks & had an operation, from which I am only slowly gaining strength. I'm afraid I'm not the man I was!

Many thanks again for your good wishes & the socks.

Yours sincerely,
W. N. Bailey.

Commentary. This is Bailey's last letter to Slater, written the day after his 68th birthday. Bailey died seven weeks later on October 23, 1961. While his illness appears to have sapped the energy necessary to continue mathematical research, his kind demeanor and positive outlook remained. Bailey makes reference to Slater's first book, *Confluent Hypergeometric Functions* [Sla60], and the reissuing of his old Cambridge tract [Bai35]. In the preface to Slater's second book [Sla66, p. xiii], *Generalized Hypergeometric Functions*, she states

> This book should really be attributed to Bailey and Slater. It was Professor Bailey's intention to write a comprehensive work on hypergeometric functions, with my assistance. This present work is based in part on notes for a series of lectures which he gave in 1947–1950 at Bedford College, London University. The rest of the book contains the results of my own researches into the general theory. It also covers the great advances made in the subject since 1936 when W. N. Bailey's Cambridge Tract 'Generalized Hypergeometric Series' was first published.

Bailey's LMS obituary [Sla62] was written by Slater. In it, we learn that

> As a lecturer and teacher he was clear and painstaking, and he was particularly good with the slower students at "making the dark places light." As a director of research he was brilliant and unsurpassed in his own field. ...He has left behind him the memory of a quiet family man, one who was an effective teacher and lecturer, one who organised his department unobtrusively but successfully, and one who inspired many of his students to carry out independent researches for themselves.

Bibliography

[AA15]   K. Alladi and G. E. Andrews, *The dual of Göllnitz's (big) partition theorem*, Ramanujan J. **36** (2015), 171–201.

[AAG95]   K. Alladi, G. E. Andrews, and B. Gordon, *Refinements and generalizations of Capparelli's conjecture on partitions*, J. Algebra **174** (1995), 636–658.

[AB87]   G. E. Andrews and R. J. Baxter, *Lattice gas generalization of the hard hexagon model III: q-trinomial coefficients*, J. Stat. Phys. **47** (1987), 297–330, reprinted in G. E. Andrews, *The Selected Works of George E. Andrews*, ed. A. V. Sills, Imperial College Press, London, 2013, pp. 343–376.

[AB89]   _____, *A motivated proof of the Rogers–Ramanujan identities*, Amer. Math. Monthly **96** (1989), 401–409.

[AB90]   _____, *Scratchpad explorations for elliptic theta functions*, Computers in Mathematics (D. V. Chudnovsky and R. D. Jenks, eds.), Lecture Notes in Pure and Applied Mathematics, vol. 125, Marcel Dekker, New York, 1990, reprinted in *The Selected Works of George E. Andrews*, ed. A. V. Sills, Imperial College Press, London, 2013, pp. 727–744, pp. 17–33.

[AB05]   G. E. Andrews and B. C. Berndt, *Ramanujan's lost notebook, part I*, Springer, 2005.

[AB09]   _____, *Ramanujan's lost notebook, part II*, Springer, 2009.

[Abb01]   Janet Abbate, *Oral history: Interview with Lucy Slater*, http://ethw.org/Oral-History:Lucy_Slater, April 2001, Oral History, Interview # 630 for the IEEE History Center, The Institute of Electrical and Electronic Engineers, Inc.

[ABF84]   G. E. Andrews, R. J. Baxter, and P. J. Forrester, *Eight-vertex SOS model and generalized Rogers–Ramanujan type identities*, J. Stat. Phys. **35** (1984), 193–266.

[ABM15]   G. E. Andrews, K. Bringmann, and K. Mahlburg, *Double series representations for Schur's partition function and related identities*, J. Combin. Theory Ser. A **132** (2015), 102–119.

[AG93]   K. Alladi and B. Gordon, *Generalizations of Schur's partition the-orem*, Manuscripta Math. **79** (1993), 113–126.

[AG95]   _____, *Schur's partition theorem, companions, refinements and generalizations*, Trans. Amer. Math. Soc. **347** (1995), no. 5, 1591–1608.

[AKP00]   G. E. Andrews, A. Knopfmacher, and P. Paule, *An infinite family of Engel expansions of Rogers–Ramanujan type*, Adv. Appl. Math. **25** (2000), 2–11.

[Ald48]   H. L. Alder, *The nonexistence of certain identities in the theory of partitions and compositions*, Bull. Amer. Math. Soc. **54** (1948), no. 8, 712–722.

[All98]   K. Alladi, *On a partition theorem of Göllnitz and quartic transfor-mations*, J. Number Theory **69** (1998), 153–180.

[And67]   G. E. Andrews, *A generalization of the Göllnitz–Gordon partition theorems*, Proc. Amer. Math. Soc. **8** (1967), 945–952.

[And68]   _____, *On q-difference equations for certain well-poised basic hy-pergeometric series*, Quart. J. Math. **19** (1968), 433–447, reprinted in G. E. Andrews, *The Selected Works of George E. Andrews*, ed. A. V. Sills, Imperial College Press, London, 2013, pp. 91–106.

[And73]   _____, *On the q-analog of Kummer's theorem and applications*, Duke Math. J. **40** (1973), no. 3, 525–528, reprinted in G. E. An-drews, *The Selected Works of George E. Andrews*, ed. A. V. Sills, Imperial College Press, London, 2013, pp. 58–61.

[And74a]   _____, *An analytic generalization of the Rogers–Ramanujan iden-tities for odd moduli*, Proc. Nat. Acad. Sci. **71** (1974), no. 10, 4082–4085, reprinted in G. E. Andrews, *The Selected Works of George E. Andrews*, ed. A. V. Sills, Imperial College Press, London, 2013, pp. 109–112.

[And74b]   _____, *On the general Rogers–Ramanujan theorem*, Memoirs Amer. Math. Soc. **152** (1974), 86 pp.

[And75]   _____, *On Rogers-Ramanujan type identities related to the mod-ulus 11*, Proc. London Math. Soc. **30** (1975), 330–346.

[And76]   _____, *The theory of partitions*, Encyclopedia of Mathematics and Its Applications, vol. 2, Addison–Wesley, Reading, MA, 1976, Reissued: Cambridge University Press, 1998.

[And79]   _____, *Partitions: Yesterday and today*, New Zealand Mathemat-ical Society, 1979.

[And81] ———, *Ramanujan's "lost" notebook ii: ϑ-function expansions*, Adv. Math. **41** (1981), 173–185.

[And82] ———, *L. J. Rogers and the Rogers–Ramanujan identities*, Math. Chronicle **11** (1982), 1–15, reprinted in G. E. Andrews, *The Selected Works of George E. Andrews*, ed. A. V. Sills, Imperial College Press, London, 2013, pp. 977–991.

[And84] ———, *Multiple series Rogers–Ramanujan type identities*, Pacific J. Math **114** (1984), 267–283, reprinted in G. E. Andrews, *The Selected Works of George E. Andrews*, ed. A. V. Sills, Imperial College Press, London, 2013, pp. 113–129.

[And86] ———, *q-series: Their development and application in analysis, number theory, combinatorics, physics, and computer algebra*, Regional Conference Series in Mathematics, vol. 66, American Mathematical Society, Providence, RI, 1986.

[And89] ———, *Physics, Ramanujan, and computer algebra*, Computer Algebra (David V. Chudnovsky and Richard D. Jenks, eds.), Marcel Dekker, Inc., New York and Basel, 1989.

[And90a] ———, *Euler's 'exemplum memorabile inductionis fallis' and q-trinomial coefficients*, J. Amer. Math. Soc. **3** (1990), 653–669, reprinted in G. E. Andrews, *The Selected Works of George E. Andrews*, ed. A. V. Sills, Imperial College Press, London, 2013, pp. 204–220.

[And90b] ———, *q-trinomial coefficients and the Rogers–Ramanujan identities*, Analytic Number Theory (B. C. Berndt et al., ed.), Birkhauser, Boston, 1990, pp. 1–11.

[And94] ———, *Schur's theorem, Capparelli's conjecture and q-trinomial coefficients*, The Rademacher Legacy to Mathematics (George E. Andrews, David M. Bressoud, and L. Alayne Parson, eds.), American Mathematical Society, 1994, The Centenary Conference in Honor of Hans Rademacher, July 21–25, 1992, The Pennsylvania State University.

[And01] ———, *Some debts I owe*, The Andrews Festschrift: Seventeen Papers on Classical Number Theory and Combinatorics (D. Foata and G.-N. Han, eds.), Springer, 2001, Also published electronically in *Séminaire Lotharingien de Combinatoire*, as paper B42a in issue 42, which consists of papers presented at a meeting in honor of George Andrews' 60th birthday held in Maratea, Italy in September 1998, pp. 1–16.

[And09] ———, *The finite Heine transformation*, Integers **9 (S)** (2009), Article 1, 6 pp.

[And13a] _____, *The discovery of Ramanujan's lost notebook*, The Legacy of Srinivasa Ramanujan (Bruce C. Berndt and Dipendra Prasad, eds.), Lecture Notes Series, vol. 20, Ramanujan Mathematical Society, India, 2013, Proceedings of an International Conference in Celebration of the 125th Anniversary of Ramanujan's birth; University of Delhi, 17–22 December 2012, pp. 77–88.

[And13b] _____, *The Selected Works of George E. Andrews (with Commentary)*, (A. V. Sills, ed.), Imperial College Press, London, 2013.

[And14] George E. Andrews, *Knots and q-series*, Ramanujan 125 (K. Alladi, F. G. Garvan, and A. J. Yee, eds.), vol. 627, American Mathematical Society, 2014, Proceedings of an International Conference to commemorate the 125th Anniversary of Ramanujan's birth; University of Florida, November 5–7, 2012, pp. 17–24.

[Apo48] T. M. Apostol, *A study of dedekind sums and their generalizations*, Ph.D. thesis, University of California at Berkeley, 1948.

[Apo76] _____, *Introduction to analytic number theory*, Undergraduate Texts in Mathematics, Springer, 1976.

[Apo97] _____, *Modular functions and Dirichlet series in number theory*, 2nd ed., Graduate Texts in Mathematics, vol. 41, Springer, 1997.

[AS97] G. E. Andrews and J. P. O. Santos, *Rogers–Ramanujan type identities for partitions with attached odd parts*, Ramanujan J. **1** (1997), 91–99.

[Ask01] R. Askey, *The work of George Andrews: a Madison perspective*, The Andrews Festschrift: Seventeen papers on classical number theory and Combinatorics, Springer, 2001, pp. 17–38.

[Bai35] W. N. Bailey, *Generalized hypergeometric series*, Cambridge Tracts in Mathematics and Mathematical Physics, Cambridge, 1935.

[Bai41] _____, *A note on certain q-identiites*, Quart. J. Math. **12** (1941), 173–175.

[Bai47] _____, *Some identities in combinatory analysis*, Proc. London Math. Soc. (2) **49** (1947), 421–435.

[Bai48] _____, *Identities of the Rogers-Ramanujan type*, Proc. London Math. Soc. (2) **50** (1948), 1–10.

[Bai51] _____, *On the simplification of some identities of the Rogers–Ramanujan type*, Proc. London Math. Soc. (3) **1** (1951), 217–221.

[Bax80] R. J. Baxter, *Hard hexagons: exact solution*, J. Phys. A **13** (1980), L61–L70.

[Bax81] ———, *Rogers–Ramanujan identities in the hard hexagon model*, J. Stat. Phys. **26** (1981), 427–452.

[Bax82] ———, *Exactly solved models in statistical mechanics*, Academic Press, 1982, Reissued, Dover, 2008.

[BF16] M. T. Batchelor and A. Foerster, *Yang–Baxter integrable models in experiments: from condensed matter to ultracold atoms*, J. Physics A **49** (2016), no. 17, 22 pp.

[BM96] A. Berkovich and B. M. McCoy, *Continued fractions and fermionic representations for characters of $M(p, p')$ minimal models*, Lett. Math. Phys. **37** (1996), 49–66.

[BM97] ———, *Generalizations of the Andrews–Bressoud identities for the $N = 1$ superconformal model $SM(2, 4\nu)$*, Math. Comput. Modelling **26** (1997), 37–49.

[BM98] ———, *Rogers–Ramanujan identities: a century of progress from mathematics to physics*, Doc. Math. (1998), 163–172, Extra volume ICM III.

[BMO96] A. Berkovich, B. M. McCoy, and W. P. Orrick, *Polynomial identities, indicies, and duality for the $N = 1$ superconformal model $SM(2, 4\nu)$*, J. Stat. Phys. **83** (1996), 795–837.

[BMS98] A. Berkovich, B. M. McCoy, and A. Schilling, *Rogers–Schur–Ramanujan type identities for the $M(p, p')$ minimal models of conformal field theory*, Comm. Math. Phys. **191** (1998), 211–223.

[BMS09] D. Bowman, J. McLaughlin, and A. V. Sills, *Some more identities of Rogers–Ramanujan type*, Ramanujan J. **18** (2009), 307–325.

[BO13] J. Bruinier and K. Ono, *Algebraic formulas for the coefficients of half-integral weight harmonic weak Maas forms*, Adv. Math. **246** (2013), 198–219.

[Bos03] M. K. Bos, *Coding the principal character formula for affine Kac–Moody Lie algebras*, Math. Comput. **72** (2003), no. 244, 2001–2012.

[BP83] R. J. Baxter and P. A. Pearce, *Hard squares with diagonal entries*, J. Phys. A **16** (1983), 2239–2255.

[BP84] ———, *Deviations from the critical density in the generalized hard hexagon model*, J. Phys. A **17** (1984), 2095–2108.

[BP06] C. Boulet and I. Pak, *A combinatorial proof of the Rogers–Ramanujan and Schur identities*, J. Combin. Theory Series A **113** (2006), 119–130.

[BR95] B. C. Berndt and R. A. Rankin, *Ramanujan: Letters and commentary*, History of Mathematics, vol. 9, American Mathematical Society and London Mathematical Society, 1995.

[Bre79] D. M. Bressoud, *A generalization of the Rogers–Ramanujan identities for all moduli*, J. Combin. Theory, Ser. A **27** (1979), 64–68.

[Bre80a] _____, *Analytic and combinatorial generalizations of the Rogers–Ramanujan identities*, Mem. Amer. Math. Soc. **227** (1980), 54 pp.

[Bre80b] _____, *A combinatorial proof of Schur's 1926 partition theorem*, Proc. Amer. Math. Soc. **1979** (1980), no. 2, 338–340.

[Bre81] _____, *Some identities for terminating q-series*, Math. Proc. Cambridge Philos. Soc. **81** (1981), 211–223.

[BZ82] D. M. Bressoud and D. Zeilberger, *A short Rogers–Ramanujan bijection*, Discrete Math. **38** (1982), 313–315.

[Cap96] S. Capparelli, *A construction of the level 3 modules for the affine algebra $A_2^{(2)}$ and a new combinatorial identity of the Rogers–Ramanujan type*, Trans. Amer. Math. Soc. **348 (2)** (1996), 481–501.

[Car59] L. Carlitz, *Some formulas related to the Rogers–Ramanujan identities*, Annali di matematica pura et applicata **47** (1959), no. 1, 243–251.

[Car05] R. Carter, *Lie algebras of finite and affine type*, Cambridge Studies in Advanced Mathematics, Cambridge University Press, 2005.

[Cay84] A. Cayley, *Note on a partition-series*, Amer. J. Math. **6** (1884), 63–64.

[CH97] H. H. Chan and S.-S. Huang, *Ramanujan–Göllnitz–Gordon continued fraction*, Ramanujan J. **1** (1997), 75–90.

[Cig04] J. Cigler, *q-Fibonacci polynomials and the Rogers–Ramanujan identities*, Ann. Combin. **8** (2004), 269–285.

[CKL+17] B. Coulson, S. Kanade, J. Lepowsky, R. McRae, Fei Qi, M. C. Russell, and C. Sadowski, *A motivated proof of the Göllnitz–Gordon–Andrews identities*, Ramanujan J. **42** (2017), 97–129.

[CL04] S. Corteel and J. Lovejoy, *Overpartitions*, Trans. Amer. Math. Soc. **356** (2004), 1623–1635.

[CL09] _____, *Overpartitions and the q-Bailey identity*, Proc. Edinburgh Math. Soc. **52** (2009), 297–306.

[Coo06]   S. Cooper, *The quintuple product identity*, Int. J. Number Theory **2** (2006), no. 1, 115–161.

[CS16]   Y. Choliy and A. V. Sills, *A formula for the partition function that 'counts'*, Ann. Combin. **20** (2016), 301–316.

[CSS12]   S. Corteel, C. Savage, and A. V. Sills, *Lecture hall sequences, q-series, and asymmetric partition identities*, Partitions, *q*-series, and Modular Forms, Developments in Mathematics, vol. 23, Springer, 2012.

[Dau42]   J. A. Daum, *The basic analogue of Kummer's theorem*, Bull. Amer. Math. Soc. **48** (1942), 711–713.

[dlVP96]   C. de la Vallé-Poussin, *Recherches analytiques sur la théorie des nombres premiers*, Ann. Soc. Sci. Bruxelles **20** (1896), 183–256, 281–297.

[DS05]   F. Diamond and J. Shurman, *A first course in modular forms*, Graduate Texts in Mathematics, vol. 228, Springer–Verlag, New York, 2005.

[Duk14]   W. Duke, *Almost a century of answering the question: What is a mock theta function?*, Notices Amer. Math. Soc. **61** (2014), 1314–1320.

[Dys43]   F. J. Dyson, *Three identities in combinatory analysis*, J. London Math. Soc. **18** (1943), 35–39.

[Dys88]   ———, *A walk through Ramanujan's garden*, Ramanujan Revisited: Proceedings of the Centenary Conference, University of Illinois at Urbana–Champaign, June 1–5, 1987 (G. E. Andrews et al., ed.), Academic Press, 1988, pp. 7–28.

[FB85]   P. J. Forrester and R. J. Baxter, *Further exact solutions of the eight vertex SOS model and generalized Rogers–Ramanujan type identities*, J. Stat. Phys. **38** (1985), 435–472.

[FOR13]   A. Folsom, K. Ono, and R. C. Rhoades, *Mock theta functions and quantum modular forms*, Forum of Mathematics, Pi **1** (2013), 27 pp.

[Fri16]   R. Fricke, *Die Elliptischen Funktionen und ihre Anwendungen*, Ersete Teil, Teubner, Leipzig, 1916.

[Gau13]   C. F. Gauss, *Disquistiones generalies circa seriem infinitam*, Comm. soc. reg. sci. Gött. rec. **2** (1813).

[GIS99]   K. C. Garrett, M. E. H. Ismail, and D. Stanton, *Variants of the Rogers–Ramanujan identities*, Adv. Appl. Math. **23** (1999), 274–299.

[GL15] S. Garoufalidis and T. T. Q. Le, *Nahm sums, stability and the colored jones polynomial*, Research Math. Sci. **2** (2015), 1–55.

[Gla83] J. W. L. Glaisher, *A theorem in partitions*, Messenger of Math. **12** (1883), 158–170.

[GM81] A. M. Garsia and S. C. Milne, *A Rogers–Ramanujan bijection*, J. Combin. Theory Series A **31** (1981), 289–339.

[Göl60] H. Göllnitz, *Einfache partitionen*, Master's thesis, Göttingen, 1960, Diplomabeit W. S., unpublished.

[Göl67] ———, *Partitionen mit Differenzenbedingungen*, J. Reine Angew. Math. **225** (1967), 154–190.

[Gor61] B. Gordon, *A combinatorial generalization of the Rogers–Ramanujan identities*, Amer. J. Math. **83** (1961), no. 2, 393–399.

[Gor65] ———, *Some continued fractions of the Rogers–Ramanujan type*, Duke Math. J. **31** (1965), 741–748.

[Gor83] ———, *Sieve-equivalence and explicit bijections*, J. Combin. Theory Series A **34** (1983), 90–92.

[GOR13] M. Griffin, K. Ono, and L. Rolen, *Ramanujan's mock theta functions*, Proc. Nat. Acad. Sci. USA **110** (2013), no. 15, 5765–5768.

[GOW16] M. Griffin, K. Ono, and S. O. Warnaar, *A framework of Rogers–Ramanujan identities and their arithmetic properties*, Duke Math. J. **165** (2016), 1475–1527.

[GP10] N. S. S. Gu and H. Prodinger, *One-parameter generalizations of Rogers–Ramanujan type identiites*, Adv. Appl. Math. **45** (2010), 149–196.

[GR04] G. Gasper and M. Rahman, *Basic hypergeometric series*, Encyclopedia of Mathematics and its Applications, vol. 96, Cambridge University Press, Cambridge, 2004.

[GS83] I. Gessel and D. Stanton, *Applications of q-Lagrange inversion to basic hypergeometric series*, Trans. Amer. Math. Soc. **277** (1983), 173–201.

[Had96] J. Hadamard, *Sur la distribution des zéros de la fonction $\zeta(s)$ et ses conséquences arithmétiques*, Bull. Soc. Math. France **24** (1896), 199–220.

[Hag63] P. Hagis, *Partitions into odd summands*, Amer. J. Math. **85** (1963), 213–222.

[Hal36] N. Hall, *An algebraic identity*, J. London Math. Soc. **11** (1936), 276.

[Har37a] G. H. Hardy, *The Indian mathematician Ramanujan*, Amer. Math. Monthly **44** (1937), no. 3, 137–155.

[Har37b] _____, *Lectures by Godfrey H. Hardy on the mathematical work of Ramanujan*, Edwards Brothers, Ann Arbor, Michigan, 1937, Fall Term 1936. Notes taken by Marshall Hall at the Institute For Advanced Study, Princeton, NJ.

[Har40] _____, *Ramanujan: Twelve lectures suggested by his life and work*, Cambridge University Press, 1940, Reissued: AMS Chelsea, 1999.

[Hei78] E. Heine, *Handbuch der Kugelfunctionen, Theorie und Anwendungen*, vol. 1, Reimer, Berlin, 1878.

[Her18] J. F. W. Herschel, *On circulating functions, and on the integration of a class of equations of finite differences into which they enter as coefficients*, Philosophical Trans. Royal Soc. **108** (1818), 144–168.

[Hic88] D. Hickerson, *A proof of the mock theta conjectures*, Invent. Math. **94** (1988), 639–660.

[HR18] G. H. Hardy and S. Ramanujan, *Asymptotic formulæ in combinatory analysis*, Proc. London Math. Soc. **17** (1918), 75–115.

[Hua42] L.-K. Hua, *On the number of partitions into unequal parts*, Trans. Amer. Math. Soc. **51** (1942), 194–201.

[Ive62] K. E. Iverson, *A programming language*, Wiley, 1962.

[Jac29] C. G. J. Jacobi, *Fundamenta nova theoriæ functionum ellipticarum*, Regiomonti, Sumtibus fractrum Bornträger, Königsberg, 1829.

[Jac28] F. H. Jackson, *Examples of a generalization of Euler's transformation for power series*, Mess. Math. **57** (1928), 169–187.

[JM85] M. Jimbo and T. Miwa, *A solvable lattice model and related Rogers–Ramanujan type identities*, Physica **15D** (1985), 335–353.

[Kac90] V. Kac, *Infinite dimensional Lie algebras*, 3rd ed., Cambridge University Press, 1990.

[Kil08] L. J. P. Kilford, *Modular forms: A classical and computational introduction*, Imperial College Press, London, 2008.

[KLRS17] S. Kanade, J. Lepowsky, M. C. Russell, and A. V. Sills, *Ghost series and a motivated proof of the Andrews–Bressoud identities*, J. Combin. Theory Ser. A **146** (2017), 33–62.

[Kno93]   M. I. Knopp, *Modular functions in analytic number theory*, second ed., AMS Chelsea, 1993.

[Knu92]   D. E. Knuth, *Two notes on notation*, Amer. Math. Monthly **99** (1992), no. 5, 403–422.

[Kob84]   N. Koblitz, *Introduction to elliptic curves and modular forms*, Graduate Texts in Mathematics, vol. 97, Springer–Verlag, New York, 1984.

[KR15]   S. Kanade and M. C. Russell, `IdentityFinder` *and some new identities of Rogers–Ramanujan type*, Experimental Mathematics **24** (2015), no. 2, 419–423.

[Leb40]   V.-A. Lebesgue, *Sommation de quelques séries*, Journal Math. Pures Appl. **5** (1840), 42–71.

[Leh41]   J. Lehner, *A partition function connected with the modulus 5*, Duke J. Math. **8** (1941), 631–655.

[Leh46]   D. H. Lehmer, *Two nonexistence theorems on partitions*, Bull. Amer. Math. Soc. **52** (1946), 538–544.

[Lep07]   J. Lepowsky, *Some developments in vertex operator algebra theory, old and new*, Lie Algebras, Vertex Operator Algebras and their Applications (Providence, RI) (Yi-Zhi Huang and Kailash C. Misra, eds.), American Mathematical Society, 2007, Proceedings of the International Conference in Honor of James Lepowsky and Robert Wilson on their Sixtieth Birthdays, May 17–21, 2005, North Carolina State University, Raleigh, North Carolina; in Contemporary Mathematics, vol. 442, pp. 355–388.

[LM78]   J. Lepowsky and S. Milne, *Lie algebraic approaches to classical partition identities*, Adv. Math. **29** (1978), 15–59.

[Lox84]   J. H. Loxton, *Special values of the dilogarithm function*, Acta Arith. **43** (1984), 155–166.

[LW78]   J. Lepowsky and R. L. Wilson, *Construction of the affine Lie algebra* $A_1^{(1)}$, Comm. Math. Phys. **62** (1978), 45–53.

[LW81]   ———, *A new family of algebras underlying the Rogers–Ramanujan identities*, Proc. Nat. Acad. Sci. USA **78** (1981), 7254–7258.

[LW82]   ———, *A Lie theoretic interpretation and proof of the Rogers–Ramanujan identities*, Adv. Math. **45** (1982), 21–72.

[LW84]   _____, *The structure of standard modules I: universal algebras and the Rogers–Ramanujan identities*, Invent. Math. **77** (1984), 199–290.

[LW85]   _____, *The structure of standard modules II: the case $A_1^{(1)}$, principal gradation*, Invent. Math. **79** (1985), 417–442.

[Lyn97]   John Lynch, *The Proof*, Nova, PBS television series, original airdate: October 28, 1997.

[LZ12]   J. Lepowsky and M. Zhu, *A motivated proof of Gordon's identities*, Ramanujan J. **29** (2012), 199–211.

[Mac18]   P. A. MacMahon, *Combinatory analysis*, vol. II, Cambridge University Press, 1918, Reprinted by the American Mathematical Society, 2001.

[Mac99]   I. G. Macdonald, *Symmetric functions and Hall polynomials*, 2nd ed., Oxford Mathematical Monographs, Oxford University Press, 1999.

[Mei54]   G. Meinardus, *Asymptotische aussagen über Partitionen*, Math. Z. **59** (1954), 388–398.

[Mel94]   E. Melzer, *Fermionic character sums and the corner transfer matrix*, Internat. J. Modern Phys. A **9** (1994), 1115–1136.

[MS08]   J. McLaughlin and A. V. Sills, *Ramanujan–Slater type identities related to the moduli 18 and 24*, J. Math. Anal. Appl. **344** (2008), 765–777.

[MS09]   _____, *Combinatorics of Ramanujan–Slater type identities*, Integers **9 (S)** (2009), Article 10, 14 pp.

[MS12]   _____, *On a pair of identities from Ramanujan's lost notebook*, Ann. Comb. **16** (2012), 591–607.

[MSZ08]   J. McLaughlin, A. V. Sills, and P. Zimmer, *Rogers–Ramanujan–Slater type identities*, Electronic J. Comb. **15** (2008), Dynamic Survey #15, 59 pp.

[MSZ09]   _____, *Rogers–Ramanujan computer searches*, J. Symbolic Computation **44** (2009), 1068–1078.

[MZ05]   M. Mohammed and D. Zeilberger, *Sharp upper bounds for the orders of the recurrences output by the Zeilberger and q-Zeilberger algorithms*, J. Symbolic Comput. **39** (2005), 201–207.

[Nan14]   D. Nandi, *Partition identities arising from the standard $A_2^{(2)}$ modules of level 4*, Ph.D. thesis, Rutgers, 2014.

[Niv40]   I. Niven, *On a certain partition function*, Amer. J. Math. (1940), 353–364.

[O'H88]   K. M. O'Hara, *Bijections for partition identities*, J. Combin. Theory Series A **49** (1988), 13–25.

[Ono03]   K. Ono, *The web of modularity: Arithmetic of the coefficients of modular forms and q-series*, CBMS Regional Conference Series in Mathematics, vol. 103, American Mathematical Society, 2003.

[Pak06]   I. Pak, *Partition bijections: a survey*, Ramanujan J. **12** (2006), 5–75.

[Pau82]   P. Paule, *Zwei neue Transformationen als elementare Anwendungen der q-Vandermonde Formel*, Ph.D. thesis, University of Vienna, 1982.

[Pau85]   _____, *On identities of the Rogers–Ramanujan type*, J. Math. Anal. Appl. **107** (1985), 255–284.

[Pau87]   P. Paule, *A note on Bailey's lemma*, J. Combin. Theory, Ser. A **44** (1987), 164–167.

[Pau94]   _____, *Short and easy computer proofs of the Rogers–Ramanujan identities and identities of similar type*, Electronic J. Combin. **1** (1994), 9 pp., # R10.

[PWZ96]   M. Petkovšek, H. S. Wilf, and D. Zeilberger, $A = B$, A. K. Peters, Ltd., Wellesley, MA, 1996.

[Rad37]   H. A. Rademacher, *On the partition function $p(n)$*, Proc. London Math. Soc. (2) **43** (1937), 241–254.

[Rad43]   _____, *On the expansion of the partition function in a series*, Ann. Math. **44** (1943), no. 3, 416–422.

[Rad73]   _____, *Topics in analytic number theory*, Springer, 1973.

[Ram27]   S. Ramanujan, *Collected papers*, (G. H. Hardy, P. V. Seshu Aiyar, and B. M. Wilson, eds.), Cambridge University Press, 1927, Reissued AMS Chelsea, 2000.

[Ram88]   _____, *The lost notebook and other unpublished papers*, Narosa Publishing House, New Delhi, Madras, Bombay, 1988.

[Rem82]   J. Remmel, *Bijective proofs of some classical partition identities*, J. Combin. Theory Series A **33** (1982), 273–286.

[Rho13]   R. C. Rhoades, *On Ramanujan's defintion of mock theta function*, Proc. Nat. Acad. Sci. USA **110** (2013), no. 19, 7592–7594.

[Rog93a]   L. J. Rogers, *Memoir on the expansion of some infinite products*, Proc. London Math. Soc. (1) **24** (1893), 337–352.

[Rog93b]   _____, *On a three-fold symmetry in the elements of Heine's series*, Proc. London Math. Soc. **24** (1893), 171–179.

[Rog94]   _____, *Second memoir on the expansion of certain infinite products*, Proc. London Math. Soc. (1) **25** (1894), 318–343.

[Rog95]   _____, *Third memoir on the expansion of certain infinite products*, Proc. London Math. Soc. (1) **26** (1895), 15–32.

[Rog17]   _____, *On two theorems of combinatory analysis and some allied identities*, Proc. London Math. Soc. **16** (1917), 315–336.

[Rot11]   H. Rothe, *Systematisches Lehrbuch der Aritmetik*, Leipzig, 1811.

[Rot09]   G.-C. Rota, *Indiscrete thoughts*, Birkhäuser, Boston, 2009.

[Sch17]   I. Schur, *Ein Beitrag zur additiven Zahlentheorie und zur Theorie der kettenbrüche*, S.-B. Preuss. Akad. Wiss. Phys. Math. Klasse (1917), 302–321.

[Sch26]   _____, *Zur additive Zahlentheorie*, S.-B. Preuss. Akad. Wiss. Phys. Math. Klasse (1926), 488–495.

[Sea52]   D. B. Sears, *Two identities of Bailey*, J. London Math. Soc. **27** (1952), 510–511.

[Sel36]   A. Selberg, *Über einige arithmetische Identitäten*, Avrandlinger Norske Akad. **8** (1936), 23 pp.

[SF82]   J. J. Sylvester and F. Franklin, *A constructive theory of partitions, arranged in three acts, an interact, and an exodion*, Amer. J. Math. **5** (1882), no. 1, 251–330.

[Sil03]   A. V. Sills, *Finite Rogers–Ramanujan type identities*, Electronic J. Combin. (2003), no. #R13, 122 pp.

[Sil04a]   _____, *q-difference equations and identities of the Rogers–Ramanujan-Bailey type*, J. Difference Equ. Appl. **10** (2004), 1069–1084.

[Sil04b]   _____, *RRtools—a Maple package for aiding the discovery and proof of finite Rogers–Ramanujan type identities*, J. Symbolic Comput. **37** (2004), 415–448.

[Sil05]   _____, *Identities of the Rogers–Ramanujan–Bailey type*, J. Math. Anal. App. **308** (2005), 669–688.

[Sil06]   _____, *On identities of the Rogers–Ramanujan type*, Ramanujan
             J. **11** (2006), 403–429.

[Sil07]   _____, *Identities of the Rogers–Ramanujan–Slater type*, Int. J.
             Number Theory **3** (2007), 293–323.

[Sla51a]   L. J. Slater, *Functions of hypergeometric type*, Ph.D. thesis, Uni-
             versity of London, Bedford College, 1951.

[Sla51b]   _____, *A new proof of Rogers's transformations of infinite series*,
             Proc. London Math. Soc. (2) **53** (1951), 460–475.

[Sla52]   _____, *Further identities of the Rogers-Ramanujan type*, Proc.
             London Math. Soc. (2) **54** (1952), 147–167.

[Sla54]   _____, *A note on equivalent product theorems*, Math. Gazette **38**
             (1954), 127–128.

[Sla60]   _____, *Confluent hypergeometric functions*, Cambridge Univer-
             sity Press, 1960.

[Sla62]   _____, *Wilfrid Norman Bailey*, J. London Math. Soc. **37** (1962),
             504–512.

[Sla66]   _____, *Generalized hypergeometric functions*, Cambridge Univer-
             sity Press, 1966.

[Smo16]   N. A. Smoot, *A partition function connected with the Göllnitz–
             Gordon identities*, Harmonic Analysis, Partial Differential Equa-
             tions, Complex Analysis, Banach Spaces, and Operator Theory:
             Celebrating Cora Sadosky's Life (M. C. Pereyra, S. Marcan-
             tongnini, A. M. Stokolos, and W. U. Romero, eds.), Association
             for Women in Mathematics, Springer, 2016.

[Sta31]   G. W. Starcher, *On identities arising from solutions to q-difference
             equations and some interpretations in number theory*, Amer. J.
             Math. **53** (1931), 801–816.

[Sta01]   D. Stanton, *The Bailey–Rogers–Ramanujan group*, q-series with
             applications in combinatorics, number theory, and physics (Provi-
             dence, RI), Contemporary Mathematics, vol. 291, American Math-
             ematical Society, 2001, Proceedings of the Conference on q-series,
             Urbana, IL, 2000.

[Ste90]   J. R. Stembridge, *Hall–Littlewood functions, plane partitions, and
             the Rogers–Ramanujan identities*, Trans. Amer. Math. Soc. **319**
             (1990), no. 2, 469–498.

[Ste07]   W. A. Stein, *Modular forms, a computational approach*, Graduate
             Studies in Mathematics, vol. 79, American Mathematical Society,
             2007.

[SW98]   A. Schilling and S. O. Warnaar, *Supernomial coefficients, polynomial identities, and q-series*, Ramanujan J. **2** (1998), 459–494.

[TM96]   J. Tannery and J. Molk, *Eléments de la théorie des fonctions elliptiques*, Gauthiers–Villars, Paris, 1896, 4 vols.

[Tod11] H. Todt, *Asymptotics in partition functions*, Ph.D. thesis, Pennsylvania State University, 2011.

[TX95]   M. Tamba and C. Xie, *Level three standard modules for $A_2^{(2)}$*, J. Pure Appl. Algebra **105 (1)** (1995), 53–92.

[Usp20] J. V. Uspensky, *Asymptotic formulæ for numerical functions which occur in the theory of partitions*, Izv. Akad. Nauk SSSR **14** (1920), 199–218, In Russian.

[VJ80]   A. Verma and V. K. Jain, *Transformations between basic hypergeometric series on different bases and identities of the Rogers–Ramanujan type*, J. Math. Anal. Appl. **76** (1980), 230–269.

[VJ82]   _____, *Transformations of non-terminating basic hypergeometric series, their contour integrals and applications to Rogers–Ramanujan type identities*, J. Math. Anal. Appl. **87** (1982), 9–44.

[Wal55] J. Wallis, *Arithmetica infinitorum*, London, 1655.

[War01a] S. O. Warnaar, *The generalized Borwein conjecture*, q-Series with Applications to Combinatorics, Number Theory, and Physics (Providence, RI), Contemporary Mathematics, vol. 291, American Mathematical Society, 2001, pp. 243–267.

[War01b] _____, *Refined q-trinomial coefficients and character identities*, J. Stat. Phys. **102** (2001), 1065–1081.

[War03] _____, *Partial theta functions I: Beyond the lost notebook*, Proc. London Math. Soc. **87** (2003), 363–395.

[Wat29] G. N. Watson, *A new proof of the Rogers–Ramanujan identities*, J. London Math. Soc. **4** (1929), 4–9.

[Wat36] _____, *The final problem: an account of the mock theta functions*, J. London Math. Soc. (1) **11** (1936), 55–80.

[Wat37] _____, *The mock theta functions (2)*, Proc. London Math Soc. (2) **42** (1937), 274–304.

[Wat44] _____, *A treatise on the theory of Bessel functions*, 2nd ed., Cambridge University Press, 1944.

[Whi26]   F. J. W. Whipple, *On well-poised series, generalized hypergeomet-ric series having parameters in pairs, each pair with the same sum*, J. London Math. Soc. **24** (1926), 247–263.

[WW27]   E. T. Whittaker and G. N. Watson, *A course of modern analysis*, 4th ed., Cambridge, 1927.

[WZ90]   H. S. Wilf and D. Zeilberger, *Rational functions certify combina-torial identities*, J. Amer. Math. Soc. **3** (1990), 147–158.

[Zag10]   D. Zagier, *Quantum modular forms*, Clay Mathematics Proceed-ings **11** (2010), 659–675.

[Zwe02]   S. Zwegers, *Mock theta functions*, Ph.D. thesis, Univ. Utrecht, the Netherlands, 2002.

[Zwe09]   _____, *On two fifth order mock theta functions*, Ramanujan J. **20** (2009), 207–214.

Index

Milton Keynes UK
Ingram Content Group UK Ltd.
UKHW040105071024
449327UK00019B/829

9 780367 657611